Fundamentals of Food Processing, Packaging, Labelling and Marketing

Fundamentals of Food Processing, Packaging, Labelling and Marketing

– Author –
Kartik Lamba

Wordpen Academics
Publishing Science That Matter

ISBN: 978-93-90438-54-9 (Hardback)

Published by : **WORDPEN ACADEMICS**
4736/23, Ansari Road, Darya Ganj,
New Delhi – 110 002
Phone: +91-011-35004336
E-mail: info@wordpenacademics.com

Printed at **:** **Replika Press Pvt. Ltd.**

Exclusive Distributor: ***Bio-Green Books, New Delhi***

Preface

Food processing is the transformation of cooked ingredients, by physical or chemical means into food, or of food into other forms. Food processing combines raw food ingredients to produce marketable food products that can be easily prepared and served by the consumer. Food processing typically involves activities such as mincing and macerating, liquefaction, emulsification, and cooking; pickling, pasteurization, and many other kinds of preservation; and canning or other packaging. Food processing include toxin removal, preservation, easing marketing and distribution tasks, and increasing food consistency. In addition, it increases yearly availability of many foods, enables transportation of delicate perishable foods across long distances and makes many kinds of foods safe to eat by de-activating spoilage and pathogenic micro-organisms. Modern supermarkets would not exist without modern food processing techniques, and long voyages would not be possible. Processed foods are usually less susceptible to early spoilage than fresh foods and are better suited for long distance transportation from the source to the consumer. When they were first introduced, some processed foods helped to alleviate food shortages and improved the overall nutrition of populations as it made many new foods available to the masses. Processing can also reduce the incidence of food borne disease. Fresh materials, such as fresh produce and raw meats, are more likely to harbour pathogenic micro-organisms capable of causing serious illnesses.

Food packaging is packaging for food. A package provides protection, tampering resistance, and special physical, chemical, or biological needs. It may bear a nutrition facts label and other information about food being offered for sale. It is critical to maintain food safety during processing, packaging, storage, logistics (including cold chain), sale, and use. Conformance to applicable regulations is mandatory. Some are country specific such as the US Food and Drug Administration and the US Department of Agriculture; others are regional such as the European Food Safety Authority. Certification programs such as the Global Food Safety Initiative are sometimes used. Food packaging considerations may include: use of hazard analysis and critical control points, verification and validation protocols, Good manufacturing practices, use of an effective quality management system, track and trace systems, and requirements for label content. Special food contact materials are used when the package is in direct contact with the food product. Depending on the packaging operation and the food, packaging machinery often needs specified daily wash-down and cleaning procedures. In deciding what type of new food products a consumer would most prefer, a manufacturer can either try to develop a new food product or try to modify or extend an existing food. For example, a sweet, flavored yogurt drink would be a new product, but milk in a new flavor (such as chocolate strawberry) would be an extension of an existing product. There are three steps to both developing and extending: generate ideas, screen ideas for feasibility, and test ideas for appeal. Only after these steps will a food product make it to national market. Of one hundred new food product ideas that are considered, only six make it to a supermarket shelf. The food industry faces numerous marketing decisions. Money can be invested in brand building (through advertising and other forms of promotion) to increase either quantity demanded or the price consumers are willing to pay for a product. Coca-Cola, for example, spends a great deal of money both on perfecting its formula and on promoting the brand. This allows Coke to charge more for its product than can makers of regional and smaller brands. Manufacturers may be able to leverage their existing brand names by developing new product lines. In profitably pricing the food, the manufacturer must keep in mind that the retailer adds approximately 50 percent to the price of a wholesale product. Promoting a food to consumers is done out of store, in store, and on package. Advertisements on television and in magazines are attempts to persuade consumers to think favorably about a product, so that they go to the store to purchase the product.

The book will be highly resourceful to researcher and regulatory agencies and those who are studying and teaching food processing.

— *Kartik Lamba*

Contents

Chapter 1

Operations of Food Services

Kitchen work and chefs The heart of the hotel is the kitchen. Guests who patronize these hotels would visit them for various reasons, for locality, comfort, pricing and even for the food that is served and presented to them. Also there are some who visit these hotels for the purpose of having a meal or snack. Whatever the case maybe, the kitchen plays an important role for the existence and sustenance of the hotel.

Working in a hotel kitchen could be interesting, tiring, challenging and even exciting. With all the modern technology and gadgetry, these kitchens have turned out be somewhat high tech and more convenient. Some kitchens in luxurious hotels are so well equipped and maintained, with freezer rooms, to separate kitchens for the preparation of cakes, bread and pastry. With all the novel stoves, ovens and microwaves it is indeed a fantastic place to work in and a wonderful experience. Depending on size and class of the hotel, the kitchen staff may vary. Most of the top hotels have an Executive Chef and working under him there are different categories of chefs.

They are Chef de Cuisine, Chef de Partie, Demi Chef, Commis Chef, Sous Chef etc. The duties of these chefs will vary from one to another. The Executive Chef or the Head Chef generally will oversee and supervise the department, and will instruct them accordingly as to what the duties for the day are. They are also responsible in all matters pertaining to the department, such as leave, attending to personal issues, attending board meetings and scheduling the work to each staff member. They are expected to work on shift basis as a hotel function 24-7. Not only preparation, the presentation of the food is also equally important. An expected attribute of a chef is creativity and novelty. They are responsible from the start of the preparation of the food to the final presentation. They are expected to present the food to the guest in a manner that is appealing. Most of the elite hotels are continuously trying to improvise ways and means to allure guest to their hotels by having food fares and promotions. This has been a successful way earning revenue and is patronized by many guests for food experiences. Being creative could help a chef to present a meal in the most appetizing way, thus winning the heart of the customer.

Good communication skills are important to be successful in your mission. While constant training of the staff is done by the Executive Chefs, they are also expected at times to speak to guests and attend to some special need. Cleanliness and good hygiene is maintained at all times, it is said that the standard of the hotel could be judged by its kitchen.

Food and Beverage Services

Food and beverage services are among the most visible locations at ski areas, and a ski area's environmental commitment is often judged by the environmental practices within these establishments. Pollution prevention (P2) opportunities for restaurants address a broad range of operations, including solid waste management, grounds and facility maintenance, and restaurant supply purchasing.

A restaurant's pre-tax profit is typically only 3 to 9 percent of its total revenue; therefore, money saved through reductions in operating costs (that is, through reduced energy consumption and water use) can significantly increase the profit margin. While some restaurants have already taken advantage of the many P2 opportunities available to them, a survey by the National Restaurant Association shows that for some of the most common P2 techniques and best practices, there is still room for improvement across the industry.

Numerous programs exist for the food and beverage service industry that help restaurants and their customers minimize environmental impacts. This chapter uses the Green Restaurant Association's approach as an outline to present the environmental practices a restaurant should consider.

Green Restaurant Association

Involvement in environmental programs is a good way for a restaurant to learn more about environmental best practice opportunities. An environmental resource for the food service industry is the Green Restaurant Association (GRA), which helps member restaurants reduce environmental impacts with twelve "eco-steps".

The GRA's primary operational components are research, environmental consulting, education, public relations and marketing, and community organizing and consumer activism.

To become a member of the GRA, a restaurant must sign a statement of its commitment to making environmental improvements based on the 12 "eco-steps" listed below.

After a member restaurant has made positive environmental changes, the GRA will include the restaurant in the "Green Restaurant Guide," which serves as a directory for environmentally conscious restaurants. The GRA also offers fee-based consulting services to member restaurants and initiates public relations and marketing initiatives, such as having restaurants featured on CNN, to increase restaurant exposure and consumer awareness.

GRA's 12 eco-steps are outlined below and discussed further in Sections 9.2 through 9.13.

1. Elimination of Polystyrene Foam (commonly known as styrofoam). Replace all polystyrene foam products with environmentally friendly alternatives: paper, bamboo or sugarcane paper, recyclable plastic, biodegradable plastic, and so on. *This is the minimum environmental standard for becoming a member of the GRA.*
2. Comprehensive Recycling. Initiate or improve recycling programs for glass, plastic, bimetal, cardboard, and mixed paper.
3. Waste Reduction and Reuse. Increase bulk purchasing and reduce excessive packaging for food, condiments, and so on. Replace disposable products with reusable alternatives: eating ware, aprons, tea strainers, cups, and so on.
4. Biodegradable Plastic. Transition to corn-based "plastic" products that are biodegradable and petroleum-free: cups, utensils, garbage bags, and straws.
5. Recycled Products. Transition to recycled products with the highest postconsumer content available and non-tree-fiber paper products: napkins, paper towels, toilet paper, office paper, take-out containers, coffee jackets, plates, and bowls.
6. Non-Chlorine-Bleached Paper Products. Transition to non-bleached or non-chlorinebleached paper products: cups, wax

paper, plates, take-out containers, bags, pastry bags and grabbers, napkins, paper towels, coffee filters, and office paper.

7. Nontoxic Cleaners, Landscaping and Pest Management. Replace hazardous chemical products with biodegradable and nontoxic alternatives; dish detergent, germicides, disinfectants, toilet bowl cleaners, drain cleaner, floor wash, floor polish, glass cleaners, degreasers, and laundry detergent. For landscaping, switch to nontoxic, nonsynthetic, and organic fertilizers, pesticides, or herbicides. For pest control, use nontoxic products or services.
8. Energy Efficiency. Improve the energy efficiency of lighting, refrigeration, air conditioning, gas appliances, and so on. Obtain assistance with using government and private rebate programs. Make connections with energy consultants to learn about more extensive programs.
9. Water Efficiency. Improve the water efficiency of toilets, faucets, laundry, sprinkler systems, and so on.
10. Composting. Divert food waste from landfills and create nutrient-rich soil for gardening and landscaping.
11. "Green" Electricity. Change to an energy provider that uses solar, wind, small-scale hydroelectric, geothermal, or methane-based power that is renewable and less polluting to ecosystems.
12. Employee Education Program. Train all employees, managers, and owners. Topics covered should include
    - An environmental profile of the restaurant industry
    - A history of environmental issues relevant to food service: landfills, water pollution, air pollution, clear-cutting, and global warming
    - Data describing the restaurant impacts (positive and negative) on the environment.

Elimination of Polystyrene Foam

Polystyrene foam (Styrofoam) is widely used in restaurants in insulated cups for hot beverages and take-out containers. Production of polystyrene involves use of known (benzene) and suspected (styrene and 1,3-butadiene) human carcinogenic substances (styrene and 1,3-butadiene).

Styrene is also known to be toxic to the reproductive system. Polystyrene can be recycled; however, its recycling rates are low.

Styrofoam is light in weight, but bulky in size, so hauling Styrofoam to the nearest available recycling facility is often not economically feasible. The value of a load of Styrofoam may not even cover the cost of shipping. To reduce environmental impacts, restaurants should consider switching from Styrofoam to non-bleached paper wraps, cardboard containers, or other sustainable food packaging.

According to a report published by the Environmental Defence Waste Reduction Task Force, McDonalds has completed the switch from polystyrene foam "clamshells" to paper-based wraps for packaging its sandwich items. The wraps provide a 70 to 90 percent reduction in packaging volume, resulting in significantly less space being consumed in landfills. Compared to the polystyrene foam boxes they replaced, the new sandwich wraps also offer a substantial savings in energy used and substantial reductions in pollutant releases measured over the full life-cycle of the packaging. McDonalds is also testing Earth Shell packaging products in 300 of its restaurants throughout the U.S.. These sustainable containers are made from potato starch, natural limestone, 100 percent postconsumer recycled fiber, biodegradable polymer and wax coatings, and water.

Comprehensive Recycling

The "Recycling Guidebook for the Hospitality and Restaurant Industry" provides a general overview of developing and implementing a comprehensive recycling program. The first step mentioned in this resource is to conduct a waste audit in order to evaluate the waste stream, enabling a restaurant to better target the commodities that should be included in a recycling program. Materials that are commonly recycled in restaurants include;

- *Paper*, including cardboard, computer paper, register tape, and telephone books
- *Metals*, including aluminum, tin, and steel cans
- *Green, brown, and clear glass*
- *#1 Polyethylene* (PET) and *#2 high-density polyethylene (HDPE)* plastics
- *Printer cartridges.*

Convenience is the key to a successful recycling program. An EPA Waste Wise tip sheet notes that a convenient collection system will encourage both customers and employees to carefully sort recyclables by material type and to eliminate contaminants. Collecting uncontaminated recyclables (commodities that are properly sorted and free of excess food and beverage waste) will save time otherwise spent in sorting out contaminants. Further, uncontaminated recyclables have higher value if they are sold. Presented below are other restaurant recycling tips listed in the "Recycling Guidebook for the Hospitality and Restaurant Industry."

- Recycling and trash bins should look different from each other and be clearly marked. Both types of bins should be conveniently located in the kitchen and bar areas so that employees will use them.
- Self-serve establishments should post signs to inform customers about the recycling program and provide specific instructions. Either strategically places bins for collection of recyclable commodities, or have customers leave such commodities on a designated counter for collection by staff.
- If lack of space is a problem, specially designed equipment such as can, glass, and plastic crushers are available to reduce the volume of recyclable materials.
- Recyclable collection bins in public areas should be well-marked. Choose bins with specialized openings, such as a hole for cans or a slot for newspapers, for these areas.
- Set up a logbook or a receipt system to record the volume of recyclables leaving the premises in order to facilitate tracking and compensation.

Waste Reduction and Reuse

Waste reduction and material reuse should be the first step in minimizing the waste that a restaurant produces because it is more efficient to reduce waste at the source or reuse material than it is to recycle. As mentioned in audit is an effective way to evaluate a restaurant's waste streams. Doing so enables staff to identify wastes that are nonessential to operations, such as excess packaging material. This process can also identify disposable products that can be replaced with durable, reusable ones. For example, disposable plastic tableware and silverware should be replaced with washable or compostable utensils.

EPA's Waste Wise program is a free, 3-year, goal-oriented program that assists businesses in assessing and reducing their waste streams.11 EPA designed Waste Wise to be a flexible program, in which the participant determines how much time and money to invest. Key aspects of successful Waste Wise programs include;

- Waste assessments
- Employee education
- Measurement and reporting
- Program maintenance.

The above aspects are fundamental in assessing, developing, implementing, and maintaining a viable waste reduction and reuse program. Waste assessments provide a full understanding of waste streams and provide a basis for targeting specific waste reduction goals. However, these goals will only be reached with the full support of the restaurant – from managers and servers to cooks and food preparation staff. Therefore, employee education is critical to the success of the program. It is also important that program efforts be measured regularly.

Tracking and documenting the costs, savings, and effects of a waste reduction and reuse program are the only ways to determine the environmental and economic impacts of implementing the program. Program data can also provide insight into the effectiveness of employee education programs. Lastly, program maintenance is imperative to realize long-term benefits from a waste reduction and reuse program. A successful program is not a result of one-time changes but a cumulative result of permanent procedural and behavioural changes. Several other guidance documents provide specific waste reduction and reuse ideas for the restaurant industry. Presented below are some examples of common waste reduction techniques that are discussed in "Food for Thought: Restaurant Guide to Waste Reduction and Recycling."

- Purchase and serve beverages in bulk rather than in bottles, cans, or individual packets.
- Buy bar mixes, juices, and coffee in bulk.
- Use health department-approved, refillable condiment containers.
- Purchase cleaning supplies in concentrated form.
- Use cloth towels and reusable table linens and tableware.
- Use vendors that take back packaging material and pallets.

Another way that restaurants can minimize their waste and overall environmental impact is purchasing products through local food supply vendors and grocers. Local sourcing is of particular benefit to the tourism sector, where visitors appreciate locally distinctive food, drink, and other products. For example, several Aspen Skiing

Company-owned restaurants purchase natural, chemical-free beef and hamburger from local ranchers to support the ranching community and thus preserve open space, as well as to provide guests with healthy food. The pilot program began in winter 1998 and expanded significantly in 2000. The beef costs twice as much per pound, however

Aspen Skiing Company absorbs the additional cost. By purchasing local produce and supplies, restaurants minimize the affects of excessive transportation and shipping. Such efforts should be accounted for when evaluating the environmental impact and overall performance of a restaurant.

Biodegradable Plastic

As long as customers wish to take their food “to go,” restaurants will need to stock disposable goods such as food containers and silverware. According to Biocorp, a manufacturer of biodegradable plastic tableware and silverware, nearly 113 billion disposable cups, 39 billion disposable eating utensils, and 29 billion disposable plates are used in the U.S. every year, and half these items are made of plastic. Because restaurants cannot control what happens to these items once they are in customers' hands, the most effective way to minimize their environmental impacts is to purchase biodegradable products.

Biodegradable products are typically made of corn, starch, or paper with an easily biodegradable coating. The Earth Shell is one example of a biodegradable food service product. Biocorp also manufactures biodegradable composting bags, silverware, plates, and beverage containers.

“Scientists Perfecting Planet-Friendly Plastics,” an article published by the Environmental News Network, explains that the starch used to create these biodegradable plastics– typically wheat gluten – costs about 15 cents per pound, whereas the least costly commercial plastics cost about $1 per pound. Thus, when this starch-based plastic becomes widely available to manufacturers, it could be the cheapest plastic available.

Currently, however, most manufactured environment-friendly plastics cost about $2.50 per pound, although recent projects have brought their costs down to approximately $1.50 per pound. Because biodegradable plastics are more expensive than regular plastic, the biodegradable plastic industry has a challenge in breaking into a very competitive market.

Recycled Products

Restaurants use several types of products that can be replaced with items made from recycled materials. Most of the products with recycled content that can easily be purchased are paper products such as napkins, toilet paper, tissues, paper towels, paper tablecloths, take-out containers, register tape, and office paper. Since 1998, EPA has required federal facilities to purchase products with recycledcontent material. The resulting increased demand for recycled content paper has driven down its cost to the point that recycled-content products are the same price or less expensive than virgin material products.

Non-chlorine-bleached Paper Products

Restaurants should avoid purchasing bleached paper products. Seventh Generation, a vendor of environmentally friendly, nontoxic consumer products, explains that chlorine is used by the paper industry for two purposes: to dissolve lignin, a natural material that holds a tree's cellulous fibers together, and to whiten the final paper product. However, chlorine reacts with the virgin natural substances of trees and recycled paper material to form both dioxins and organochlorines.

As these substances are discharged and accumulate in the environment, they can have profound long-and short-term health effects on exposed humans and wildlife. One concern regarding dioxins is that they are known carcinogens, or cancer-causing substances. Short-term reactions to overexposure to chlorine may also occur, including airway inflammation and bronchial hyperresponsiveness. Organochlorines are of concern because they are suspected to be endocrine modifiers, which act as hormones in the body and can disrupt the human immune system.

Nontoxic Cleaners, Landscaping, and Pest Management

Restaurant personnel use a wide variety of chemicals to clean kitchen, dishwashing, and restroom facilities. Some cleaning products used include toilet and tile cleaner, glass cleaner, carpet cleaner, spot remover, disinfectant, and oven cleaner. Many of these products contain chemicals that are harmful to human health and the environment, which is a concern for both restaurant staff and customers. Vendors offer environmentally preferable cleaning supplies with equal or better cleaning performance at equal or less cost. Many restaurants have outdoor areas with landscaping. There are several relatively simple steps that a restaurant manager can take to minimize the environmental impact of maintaining these areas, including water conservation and selection of climate-appropriate indigenous plants. Some "green" practices for landscape management include

- Watering vegetation using "deficit" irrigation, or frequent light watering
- Selecting plants based on watering needs (Typically, indigenous plants are most efficient in using water.)
- Watering grounds during the coolest hours of the day (typically at night) Finally, pest management can also be an issue for restaurants.

Energy Efficiency

Most restaurant operations, such as cooking and cleaning, are energy-intensive. EPA's EnergyStar® Program. estimates that reducing energy consumption by 20 percent can increase a restaurant's profit by one-third, a good economic incentive to make energy efficiency investments. Making this sort of reduction is feasible when restaurants implement strategic energy conservation measures. The EnergyStar® Program analysis of restaurant energy use shows that over 60 percent of a restaurant's energy is used for cooking, heating water, and heating the establishment.

The National Restaurant Association also published tips for reducing energy costs for restaurants.

Turn on and Off

- Preheat equipment in accordance with manufacturer specifications; post preheating times near equipment.
- Turn off appliances when not in use.
- Install occupancy sensors in walk-in refrigerators and freezers.

Cooking Efficiency – Lower it/Fill it up

- Cook using equipment at full capacity when possible.
- Cook at lowest temperatures first.
- Turn off equipment during downtime.
- Use lids to minimize heat loss.

Watch Thermostats – Stay Cool and Save Money

- Set thermostats to manufacturer-recommended temperatures.
- Apply new-generation "clear" coatings to reduce solar gain from large, south-and west-facing windows.
- Use the "unoccupied" and "night setback" thermostat options.

- Ensure tamper-proof temperature settings by using locking covers on thermostats.

Keep It Clean for Energy Efficiency

- Clean condenser and evaporator coils on air conditioning and refrigeration equipment.
- Change all filters regularly.
- Preventively maintain equipment.

Dishwashing Equipment

- Heat water only to the temperature required for specific tasks.
- Install equipment of proper size.
- Fully load the machine for each cycle.

As restaurants undergo renovation or technology upgrades, managers should evaluate the energy efficiency of new equipment. Green Seal is an independent, nonprofit organization dedicated to protecting the environment by promoting the manufacture and sale of environmentally responsible consumer products.

Green Seal sets environmental standards and awards a "Green Seal of Approva" to products that cause less harm to the environment than other, similar products. Commercial consumers can use Green Seal's web site to search for resourceefficient Products

Water Efficiency

Restaurants use water in almost every aspect of their operations, including food preparation and cooking, cleaning, in restroom facilities, and as a beverage for customers. For each of these aspects, there are technologies or best management practices (BMP) that conserve water. In a collaborative effort, Agricultural and Biological Engineering, EPA, and Purdue University created an environmental enrichment toolkit for the lodging industry. Included in this document is a section that outlines the following water conservation tactics for restaurants:

- Wash food products in buckets, bowls, or other containers.
- Only run dishwashers with full loads.
- Regularly inspect dishwasher pumps for water leaks.
- Defrost or thaw frozen foods in the refrigerator instead of water.
- Install low-flow taps in kitchens and restrooms.
- Use low-flow toilets in restrooms.

- Immediately fix any leaking or dripping faucet.
- Install infrared-activated faucets and toilets in restrooms.
- Purchase and use water-saving kitchen equipment.
- Track water consumption by regularly monitoring utility bills.
- Establish an effective employee training program on water conservation.

Composting

EPA estimates that food wastes comprise 6.7 percent by weight of the total U.S. municipal solid waste stream. One option for diverting food wastes from landfills is composting. Businesses with well-established composting programs divert 50 to 100 percent of their food scraps and reduce their overall solid waste by 33 to 85 percent.

Composting can be done both on and off site; however, for restaurants where space is limited, the most feasible option is to collect food scraps for an off-site composting program. This option also transfers the responsibility of monitoring the chemical balance of the compost from the restaurant owner to the local composting operation.

However, technologies are available that enable restaurants to manage their compost on site.

Case Study: Composting At Keystone Ranch

Keystone Ranch, a horse ranch resort in Keystone, Colorado, implemented a composting program that combines horse manure with vegetable and meat scraps from the Ranch Restaurant.

The ranch rents a 30-cubic yard container from a local hauler that holds almost 20 tons of material. Biodegradable cornstarch-based bags are placed at four stations to collect food waste during food preparation and uneaten food from customers. Keystone Ranch estimates that between 10 and 15 percent of the contents of the full 30-cubic yard container is food waste. The rest of the compost mixture is manure collected from the stalls of 15 to 80 horses.

The hauler transports the container to the Twin Landfill Corporation in a nearby town, where the compost is further processed, screened, and then sold to landscapers and farmers. As a result of the composting program, which complements an established glass, aluminum, tin, cardboard, and paper recycling program, the Ranch Restaurant is "rapidly approaching a zero-waste status." Because the tipping fee for compostable material is 30 percent less than that for solid waste, the program sustains itself financially and has diverted 450 tons of organic waste from the local landfill.

Green Electricity

In addition to efforts to minimize their energy use, restaurants can minimize the environmental impacts of the energy they do use by purchasing renewable electricity. By using a renewable source of energy, a restaurant decreases its dependency on nonrenewable fossil fuels and eliminates greenhouse gas emissions associated with its electricity use. Renewable energy sources include wind, solar, geothermal, and biomass.

The availability of green electricity varies from region to region, so restaurants interested in purchasing energy from a renewable source should contact their city or state government agencies or local electric utility. Non-profit green energy certification programs, such as Green-e and Cleaner and Greener, may also be able to help direct restaurant owners to local green energy providers.

Employee Education Program

The development of an employee education program is critical to maintaining an environmentally responsible restaurant. An environmental education program should teach both the "how" and "why" of each aspect of the program. An understanding of why an employee should ensure that certain environmental procedures are followed adds meaning to the task. Because of high turnover in many restaurants, environmental education should be a regular, multimedia program, and should include both written and verbal instruction in all languages of the employees. Environmental awareness expectations should be outlined during new-hire training and reinforced throughout the period of employment.

Restaurant managers should consider acknowledging employees who demonstrate superior dedication to minimizing environmental impacts. For example, employees may be offered a percentage of the cost savings resulting from environmental projects implemented at their suggestion. Such recognition would not only reward the employees, but would send a message to customers that the restaurant is actively pursuing "greener" practices.

Once the expectation of excellent environmental performance is established, maintaining the program will require less effort. Finally, restaurant managers should encourage employees to provide feedback regarding areas for improvement. Many employees will have ideas of how to improve the environment they work in, and with a forum to voice ideas, innovations can be made.

Food and Beverage Serving and Related Workers

Nature of the Work

Food and beverage serving and related workers are the front line of customer service in full-service restaurants, casual dining eateries, and other food service establishments. These workers greet customers, escort them to seats and hand them menus, take food and drink orders, and serve food and beverages. They also answer questions, explain menu items and specials, and keep tables and dining areas clean and set for new diners. Most work as part of a team, helping coworkers to improve workflow and customer service.

Waiters and waitresses, also called *servers*, are the largest group of these workers. They take customers' orders, serve food and beverages, prepare itemized checks, and sometimes accept payment. Their specific duties vary considerably, depending on the establishment. In casual-dining restaurants serving routine, straightforward fare, such as salads, soups, and sandwiches, servers are expected to provide fast, efficient, and courteous service. In fine dining restaurants, where more complicated meals are prepared and often served over several courses, waiters and waitresses provide more formal service emphasizing personal, attentive treatment at a more leisurely pace.

Waiters and waitresses may meet with managers and chefs before each shift to discuss the menu and any new items or specials, review ingredients for potential food allergies, or talk about any food safety concerns. They also discuss coordination between the kitchen and the dining room and any customer service issues from the previous day or shift. In addition, waiters and waitresses usually check the identification of patrons to ensure they meet the minimum age requirement for the purchase of alcohol and tobacco products wherever those items are sold.

Waiters and waitresses sometimes perform the duties of other food and beverage service workers, including escorting guests to tables, serving customers seated at counters, clearing and setting up tables, or operating a cash register. However, full-service restaurants frequently hire other staff, such as hosts and hostesses, cashiers, or dining room attendants, to perform these duties.

Bartenders fill drink orders either taken directly from patrons at the bar or through waiters and waitresses who place drink orders for dining room customers. Bartenders check the identification of customers seated at the bar to ensure they meet the minimum age requirement for the purchase of alcohol and tobacco products. They prepare mixed drinks, serve bottled or draught beer, and pour wine or other beverages. Bartenders must know a wide range of drink recipes and be able to mix drinks accurately, quickly, and without waste. Some establishments, especially those with higher volume, use equipment that automatically measures, pours, and mixes drinks at the push of a button. Bartenders who use this equipment, however, still must work quickly to handle a large volume of drink orders and be familiar with the ingredients for special drink requests. Much of a bartender's work still must be done by hand.

Besides mixing and serving drinks, bartenders stock and prepare garnishes for drinks; maintain an adequate supply of ice, glasses, and other bar supplies; and keep the bar area clean for customers. They also may collect payment, operate the cash register, wash glassware and utensils, and serve food to customers who dine at the bar. Bartenders usually are responsible for ordering and maintaining an inventory of liquor, mixers, and other bar supplies.

Hosts and hostesses welcome guests and maintain reservation and waiting lists. They may direct patrons to coatrooms, restrooms, or to a place to wait until their table is ready. Hosts and hostesses assign guests to tables suitable for the size of their group, escort patrons to their seats, and provide menus. They also enter reservations, arrange parties, and assist with other special requests. In some restaurants, they act as cashiers.

Dining room and cafeteria attendants and bartender helpers—sometimes referred to collectively as the bus staff—assist waiters, waitresses, and bartenders by cleaning and setting tables, removing dirty dishes, and keeping serving areas stocked with supplies.

They may also assist waiters and waitresses by bringing meals out of the kitchen, distributing dishes to individual diners, filling water glasses, and delivering condiments. They may carry trays to dining tables for patrons. *Bartender helpers* keep bar equipment clean and glasses washed. *Dishwashers* clean dishes, cutlery, and kitchen utensils and equipment.

Food also is prepared and served in limited-service eateries, which don't employ servers and specialize in simpler preparations that often are made in advance. Two occupations with large numbers of workers are common in these types of establishments: *combined food preparation and serving workers, including fast food*; and *counter attendants, cafeteria, food concession, and coffee shop*. Combined food preparation and serving workers are employed primarily by fast food restaurants. They take food and beverage orders, retrieve items when ready, fill drink cups, and accept payment.

They also may heat food items and assemble salads and sandwiches, which constitutes food preparation. Counter attendants take orders and serve food in snack bars, cafeterias, movie theatres, and coffee shops over a counter or steam table. They may fill cups with coffee, soda, and other beverages and may prepare fountain specialties, such as milkshakes and ice cream sundaes. Counter attendants take carryout orders from diners and wrap or place items in containers.

They clean counters, write itemized bills, and sometimes accept payment. Other workers, referred to as *foodservers, nonrestaurant*, serve food to patrons outside of a restaurant environment. They might deliver room service meals in hotels or meals to hospital rooms or act as carhops, bringing orders to parked cars.

Work environment. Food and beverage service workers are on their feet most of the time and often carry heavy trays of food, dishes, and glassware. During busy dining periods, they are under pressure to serve customers quickly and efficiently. The work is relatively safe, but injuries from slips, cuts, and burns often result from hurrying or mishandling sharp tools. Three occupations—food servers, nonrestaurant; dining room and cafeteria attendants and bartender helpers; and dishwashers—reported higher incident rates than many occupations throughout the economy.

Part-time work is more common among food and beverage serving and related workers than among workers in almost any other occupation. In 2008, those on part-time schedules included half of all waiters and waitresses and almost three-fourths of all hosts and hostesses.

Food service and drinking establishments typically maintain long dining hours and offer flexible and varied work opportunities. Many food and beverage serving and related workers work evenings, weekends, and holidays. The long business hours allow for more flexible schedules that appeal to many teenagers who can gain valuable work experience. More than one-fifth of all food and beverage serving and related workers were 16 to 19 years old in 2008—about six times the proportion for all workers.

Training, Other Qualifications, and Advancement

Most food and beverage service jobs are entry level and require a high school diploma or less. Generally, training is received on the job; however, those who wish to work at more upscale restaurants, where income from tips is greater and service standards are higher, may need previous experience or vocational training.

Education and training. There are no specific educational requirements for most food and beverage service jobs. Many employers prefer to hire high school graduates for waiter and waitress, bartender, and host and hostess positions, but completion of high school usually is not required for fast-food workers, counter attendants, dishwashers, and dining room attendants and bartender helpers.

Many entrants to these jobs are in their late teens or early twenties and have a high school education or less. Usually, they have little or no work experience. Food and beverage service jobs are a major source of part-time employment for high school and college students, multiple job holders, and those seeking supplemental incomes.

All new employees receive some training from their employer. They learn safe food handling procedures and sanitation practices, for example. Some employers, particularly those in fast-food restaurants, teach new workers using self-study programs, on-line programs, audiovisual presentations, and instructional booklets that explain food preparation and service skills. But most food and beverage serving and related workers pick up their skills by observing and working with more experienced workers.

Some full-service restaurants also provide new dining room employees with some form of classroom training that alternates with periods of on-the-job work experience.These training programs communicate the operating philosophy of the restaurant, help establish a personal rapport with other staff, teach formal serving techniques, and instill a desire to work as a team. They also provide an opportunity to discuss customer service situations and the proper ways to handle unpleasant circumstances or unruly patrons.

Some food serving workers can acquire more skills by attending relevant classes offered by public or private vocational schools, restaurant associations, or large restaurant chains. Some bartenders acquire their skills through formal vocational training either by attending a school for bartending or a vocational and technical school where bartending classes are taught. These programs often include instruction on State and local laws and regulations, cocktail recipes, proper attire and conduct, and stocking a bar. Some of these schools help their graduates find jobs.

Although few employers require any minimum level of educational attainment, some specialized training is usually needed in food handling and legal issues surrounding serving alcoholic beverages. Employers are more likely to hire and promote employees based on people skills and personal qualities than education.

Other qualifications: Restaurants rely on good food and customer service to retain loyal customers and succeed in a competitive industry. Food and beverage serving and related workers who exhibit excellent personal qualities—such as a neat appearance, an ability to work as part of a team, and a natural rapport with customers—will be highly sought after. Most States require workers who serve alcoholic beverages to be at least 18 years of age, but some States require servers to be older. For bartender jobs, many employers prefer to hire people who are 25 or older. All servers that serve alcohol need to be familiar with State and local laws concerning the sale of alcoholic beverages.

Waiters and waitresses need a good memory to avoid confusing customers' orders and to recall faces, names, and preferences of frequent patrons. Knowledge of a foreign language can be helpful to communicate with a diverse clientele and staff.

Restaurants and hotels that have rigid table service standards often offer higher wages and have greater income potential from tips, but they may also have stiffer employment requirements, such as prior table service experience or higher education attainment than other establishments. Advancement. Due to the relatively small size of most food-serving establishments, opportunities for promotion are limited. After gaining experience, some dining room and cafeteria attendants and bartender helpers advance to waiter, waitress, or bartender jobs.

For waiters, waitresses, and bartenders, advancement usually is limited to finding a job in a busier or more expensive restaurant or bar where prospects for tip earnings are better. Some bartenders, hosts and hostesses, and waiters and waitresses advance to supervisory jobs, such as dining room supervisor, maitre d', assistant manager, or restaurant general manager.

A few bartenders open their own businesses. In larger restaurant chains, food and beverage service workers who excel often are invited to enter the company's formal management training program.

Employment

Food and beverage serving and related workers held 7.7 million jobs in 2008. The distribution of jobs among the various food and beverage serving occupations was as follows:

Combined food preparation and serving	
workers, including fast food	2,701,700
Waiters and waitresses	2,381,600
Counter attendants, cafeteria,	
food concession, and coffee shop	525,400
Dishwashers	522,900
Bartenders	508,700
Dining room and cafeteria attendants	
and bartender helpers	420,700
Hosts and hostesses, restaurant,	
lounge, and coffee shop	350,700
Food servers, nonrestaurant	189,800
All other food preparation and	
serving related workers	50,900

The overwhelming majority of jobs for food and beverage serving and related workers were found in food services and drinking places, such as restaurants, fast food outlets, bars, and catering or contract food service operations. Other jobs were in hotels, motels, and other traveller accommodation establishments; amusement, gambling, and recreation establishments; educational services; nursing care facilities; and civic and social organizations.

Jobs are located throughout the country but are more plentiful in larger cities and tourist areas. Vacation resorts offer seasonal employment.

Job Outlook

Average employment growth is expected, and job opportunities should be excellent for food and beverage serving and related workers as turnover is generally very high among these workers, but job competition is often keen for jobs at upscale restaurants.

Employment change. Overall employment of these workers is expected to increase by 10 percent over the 2008-18 decade, which is about as fast as the average for all occupations. Food and beverage serving and related workers are projected to have one of the largest numbers of new jobs arise, about 761,000, over this period. The growth in jobs is expected to increase as the population continues to expand. However, employment will grow more slowly than in the past as people change their dining habits. The growing popularity of take-out food and the growing number and variety of places that offer carryout options, including at many full-service restaurants, will slow the growth of waiters and waitresses and other serving workers.

Projected employment growth will vary by job type. Employment of combinedfood preparation and serving workers, which includes fast-food workers, is expected to increase faster than the average for all occupations. The limited service segment of the food services and drinking places industry has a low price advantage, fast service, and has been adding healthier foods. Slower than average employment growth is expected for waiters and waitresses, hosts and hostesses, and dining room and cafeteria attendants and bartender helpers, as more people use take-out service.

Employment of bartenders, dishwashers, and counter attendants, cafeteria, food concession, and coffee shop will grow about as fast as average. Nonrestaurant servers, such as those who deliver food trays in hotels, hospitals, residential care facilities, or catered events, are expected to have average employment growth.

Job prospects. Job opportunities at most eating and drinking places will be excellent because many people in these occupations change jobs frequently, which creates a large number of openings. Keen competition is expected, however, for jobs in popular restaurants and fine dining establishments, where potential earnings from tips are greatest.

Other Customer Services

Customer satisfaction Studies have shown that te overall satisfaction of the guest in a particular hotel determines the likelihood of the client returning to same. The study identifies that the likely influences of the guest or customer satisfaction being room quality and comfort, staff services, security and amenities topping the priority list. Customer satisfaction is very important for a hotel to survive. It is important to give prompt attention to the needs of the customer without delay. This could where a reservation or booking is concerned or even a room in a particular floor and facilities required, giving proper feed back and seeing that it is done at the time the guest checks in is very important.

It could be very annoying for the guest to find out that his/her needs have not been met, and the guest could be put off about the hotel even before he enters. As such, delivering what you have promised, will gain the confidence of the customer.

An important factor in the hospitality industry, but sometimes goes unheeded is being friendly and approachable. The customer should be made comfortable, and they should be treated as if he is the most important guest in the hotel. A friendly and courteous approach will make the customer feel welcomed. As such at all times remaining polite is very important in the hospitality industry.

Maintaining customer service policies are priority. Do not treat one guest different to the other. In case, there are issues have clear rules, where the guest is not sent from pillar to post to sort out issue. Intervene, take the matter upon yourself and resolve so that the customer is aware that no matter what, the hotel will find a solution.

Customers feel important and recognized, when the hotel goes out that extra mile to attend to dtail. Sending out personalized letter, rather than the normal standard hotel leaflets is a sure way to satisfy the customer. Make note of birthdays and anniversaries and other important events during their stay with you. If you had paid attention to detail, you would have these facts with you, or even at the time the reservation is made, the guest would have mentioned it to you. Complimentary notes, flowers or chocolates to the guest could ensure satisfaction. Anticipate your clients need, let it be a book, software, CD or even a requirement, making note of same and supplying it free of charge to the guest, will make the guest feel happy and content.

Some of these will be time consuming but cost effective, some will require just commitment. The existence of the hotel depends on the staff and the management and their approach to the needs and the wants of the guest. As such, taking note and making sure that the guest is satisfied at all cost, will in return bring success and revenue to the hotel.

Make note that one satisfied customer is adequate to promote the hotel to another.

Cashiers

Supermarkets, department stores, gasoline service stations, movie theaters, restaurants, and many other businesses employ cashiers to register the sale of their goods and services. Although specific job duties vary by employer, cashiers usually are assigned to a register at the beginning of their shifts and are given a drawer containing a specific amount of money with which to start—their "till." They must count their till to ensure that it contains the correct amount of money and adequate supplies of change.

Some cashiers also handle returns and exchanges. When they do, they must ensure that returned merchandise is in good condition, and determine where and when it was purchased and what type of payment was used.

After entering charges for all items and subtracting the value of any coupons or special discounts, cashiers total the customer's bill and take payment. Forms of payment include cash, personal checks, and gift, credit, and debit cards. Cashiers must know the store's policies and procedures for each type of payment the store accepts.

For checks and credit and debit card charges, they may request additional identification from the customer or call in for an authorization. They must verify the age of customers purchasing alcohol or tobacco. When the sale is complete, cashiers issue a receipt to the customer and return the appropriate change. They may also wrap or bag the purchase.

At the end of their shifts, cashiers once again count the drawers' contents and compare the totals with sales data. An occasional shortage of small amounts may be overlooked but, in many establishments, repeated shortages are grounds for dismissal. In addition to counting the contents of their drawers at the end of their shifts, cashiers usually separate and total charge forms, return slips, coupons, and any other noncash items.

Most cashiers use scanners and computers, but some establishments still require price and product information to be entered manually. In a store with scanners, a cashier passes a product's Universal Product Code over the scanning device, which transmits the code number to a computer. The computer identifies the item and its price. In other establishments, cashiers manually enter codes into computers and then descriptions of the items and their prices appear on the screen.

Depending on the type of establishment, cashiers may have other duties as well. In many supermarkets, for example, cashiers weigh produce and bulk food, as well as return unwanted items to the shelves. In convenience stores, cashiers may be required to know how to use a variety of machines other than cash registers, and how to furnish money orders and sell lottery tickets. Operating ticket-dispensing machines and answering customers' questions are common duties for cashiers who work at movie theaters and ticket agencies.

Work environment. Most cashiers work indoors, usually standing in booths or behind counters. Often, they are not allowed to leave their workstations without supervisory approval because they are responsible for large sums of money. The work of cashiers can be very repetitious, but improvements in workstation design in many stores are alleviating problems caused by repetitive motion. In addition, the work can sometimes be dangerous; the risk from robberies and homicides is much higher for cashiers than for other workers, although more safety precautions are being taken to help deter robbers.

About 47 percent of all cashiers worked part time in 2008. Hours of work often vary depending on the needs of the employer. Generally, cashiers are expected to work weekends, evenings, and holidays to accommodate customers' needs. However, many employers offer flexible schedules. Because the holiday season is the busiest time for most retailers, many employers restrict the use of vacation time from Thanksgiving through the beginning of January.

Training, other Qualifications, and Advancement

Cashier jobs usually are entry-level positions requiring little or no previous work experience. They require good customer service skills. Education and training. Although there are no specific educational requirements, employers filling full-time jobs often prefer applicants with high school diplomas.

Nearly all cashiers are trained on the job. In small businesses, an experienced worker often trains beginners. The trainee spends the first day observing the operation and becoming familiar with the store's equipment, policies, and procedures. After this, trainees are assigned to a register—frequently under the supervision of an experienced worker. In larger businesses, trainees spend several days in classes before being placed at cash registers. Topics typically covered in class include a description of the industry and the company, store policies and procedures, equipment operation, and security.

Training for experienced workers is not common, except when new equipment is introduced or when procedures change. In these cases, the employer or a representative of the equipment manufacturer trains workers on the job. Other qualifications. People who want to become cashiers should be able to do repetitious work accurately. They also need basic mathematics skills and good manual dexterity. Because cashiers deal constantly with the public, they should be neat in appearance and able to deal tactfully and pleasantly with customers. In addition, some businesses prefer to hire workers who can operate specialized equipment or who have business experience, such as typing, selling, or handling money.

Advancement. Advancement opportunities for cashiers vary. For those working part time, promotion may be to a full-time position. Others advance to head cashier or cash-office clerk. In addition, this job offers a good opportunity to learn about an employer's business and can serve as a steppingstone to a more responsible position.

Employment

Cashiers held about 3.55 million jobs in 2008. Although cashiers are employed in almost every industry, 24 percent of all jobs were in grocery stores. Gasoline stations, department stores, and other retail establishments also employed large numbers of these workers. Outside of retail establishments, many cashiers worked in food services and drinking places.

Job Outlook

Cashiers are expected to grow more slowly than the average for all occupations. Opportunities for full-time and part-time jobs are expected to be good because of the need to replace the large number of workers who leave this occupation.

Employment change. Employment of cashiers is expected to grow by 4 percent between 2008 and 2018 which is slower than the average for all occupations. Continued growth in retail sales is expected, but the rising popularity of purchasing goods online will limit the employment growth of cashiers, although many customers still prefer the traditional method of purchasing goods at stores. Also, the growing use of self-service checkout systems in retail trade, especially at grocery stores, should have an adverse effect on employment of cashiers. These self-checkout systems may outnumber checkouts with cashiers in the future in many establishments.

The impact on job growth for cashiers will largely depend on the public's acceptance of this self-service technology. Job prospects. Opportunities for full-time and part-time cashier jobs should continue to be good because of the need to replace the large number of workers who transfer to other occupations or leave the labour force. There is substantial movement into and out of the occupation because education and training requirements are minimal and the predominance of part-time jobs is attractive to people seeking a short-term source of income rather than a full-time career. Historically, workers under the age of 25 have filled many of the openings in this occupation. In 2008, about 47 percent of all cashiers were 24 years of age or younger.

Because cashiers are needed in businesses and organizations of all types and sizes, job opportunities are found throughout the country. However, job opportunities may vary from year to year because the strength of the economy affects demand for cashiers. Companies tend to hire more cashiers when the economy is strong. Seasonal demand for cashiers also causes fluctuations in employment.

Chefs, Head Cooks, andFood Preparation and Serving Supervisors

Chefs, head cooks, andfood preparation and serving supervisors oversee the daily food service operation of a restaurant or other food service establishment. *Chefs and head cooks* are usually responsible for directing cooks in the kitchen, dealing with food-related concerns, and providing leadership. They are also the most skilled cooks in the kitchen and use their creativity and knowledge of food to develop and prepare recipes. *Food preparation and serving supervisors* oversee the kitchen and non-kitchen staff in a restaurant or food service facility. They may also overseefood preparation workers in fast food, cafeteria, or casual dining restaurants, whre the menu is fairly standard from day to day, or in more formal restaurants, where a chef provides specific guidelines and exacting standards on how to prepare each item.

All of these workers—chefs, head cooks, and food preparation and serving supervisors—hire, train, and supervise staff, prepare cost estimates for food and supplies, set work schedules, order supplies, and ensure that the food service establishment runs efficiently and profitably. Additionally, these workers ensure that sanitation and safety standards are observed and comply with local regulations. Fresh food must be stored and cooked properly, work surfaces and dishes clean and sanitary, and staff and customers safe from illness or injury to avoid being closed by the health department or law enforcement.

While all chefs have a role in preparing the food, developing recipes, determining serving sizes, planning menus, ordering food supplies, and overseeing kitchen operations to ensure uniform quality and presentation of meals, different types of chefs may have unique roles to perform or specialize in certain aspects of the job. *Executive chefs, head cooks, and chefs de cuisine,* are primarily responsible for coordinating the work of the cooks and directing the preparation of meals. Executive chefs are in charge of all food service operations and also may supervise several kitchens of a hotel, restaurant or corporate dining operation. A *sous chef*, or sub chef, is the second-in-command and runs the kitchen in the absence of the chef. Many chefs earn fame both for themselves and for their kitchens because of the quality and distinctive nature of the food they serve.

As a greater variety of establishments prepare and serve food, chefs and head cooks and first-line supervisors of food preparation and serving workers can be found in a greater variety of places. Grocery and specialty food stores employ these workers to develop recipes and prepare meals for customers to carry out. They increasingly work in residential care facilities, such as nursing homes, and in schools and hospitals.

Some chefs and head cooks work for individuals rather than for restaurants, cafeterias, or food manufacturers. *Personal chefs* and *private household cooks* plan and prepare meals in private homes according to the client's tastes or dietary needs. They order groceries and supplies, clean the kitchen, and wash dishes and utensils. They also may serve meals. Personal chefs usually prepare a week's worth of meals in the client's home for the client to heat and serve according to directions. They may be self-employed or work as part of a team of personal chefs and employed by a company that provides this service. Private household cooks typically work full time for one client, such as corporate executives, university presidents, or diplomats, who regularly entertain as a part of their official duties.

While the work of chefs and head cooks is concentrated in the kitchen or in providing overall guidance, food preparation and serving supervisors oversee specific areas of operation in food service establishments or the kitchen and counter areas of quick service restaurants. In fast food and casual dining restaurants, they may share many of the same functions with food service managers. They are responsible for dealing with customer complaints, balancing the books at the end of the day, scheduling workers, and ordering supplies. They also supervise and train kitchen and food preparation staff and ensure that these workers know how to gather food supplies, operate equipment, and assemble orders.

Work environment. Restaurants and other food service facilities where these workers are employed are required to be clean and sanitary. Although the seating areas of eating places are often attractive, kitchens can be crowded and hot and filled with potential dangers, such as hot ovens and slippery floors. Job hazards for those working in kitchens include slips and falls, cuts, and burns, but these injuries are seldom serious. Chefs, head cooks, and supervisors are under constant pressure to get meals prepared quickly, while ensuring quality is maintained and safety and sanitation guidelines are observed. Because the pace can be hectic during peak dining times, workers must be able to communicate clearly so that food orders and service are done correctly.

Work hours in restaurants may include early mornings, late evenings, holidays, and weekends. Schedules for those working in offices, factories and school cafeterias may be more regular. In fine-dining restaurants, work schedules tend to be longer because of the time required to prepare ingredients in advance. Many executive chefs regularly work 12-hour days because they oversee the delivery of foodstuffs early in the day, plan the menu, and prepare those menu items that require the most skill. Depending upon the days of operation, some chefs or other supervisors may take less busy days off to offset the longer hours on other days.

Training, Other Qualifications, and Advancement

Most workers in these occupations have prior experience in the food service or hospitality industries. Most start as food preparation workers or line cooks in a full-service restaurant and work their way up to positions with more responsibility. Some attend cooking school or take vocational training classes and participate in internships or apprenticeship programs to acquire the additional skills needed to create menus and run a business.

Education and training. While most chefs, head cooks, and food preparation and serving supervisors have some postsecondary training, many experienced workers with less education can still be promoted. Formal training may take place at a community college, technical school, culinary arts school, or a 2-year or 4-year college with a degree in hospitality. A growing number of chefs participate in training programs sponsored by independent cooking schools, professional culinary institutes, 2-year or 4-year colleges with a hospitality or culinary arts department, or in the armed forces. Some large hotels and restaurants also operate their own training and job-placement programs for chefs and head cooks. Executive chefs, head cooks, and sous chefs who work in fine-dining restaurants require many years of training and experience.

For students in culinary training programs, most of their time is spent in kitchens learning to prepare meals by practicing cooking skills. They learn knife techniques and proper use and care of kitchen equipment. Training programs also include courses in nutrition, menu planning, portion control, purchasing and inventory methods, proper food storage procedures, and use of leftover food to minimize waste. Students also learn sanitation and public health rules for handling food. Training in food service management, computer accounting and inventory software, and banquet service are featured in some training programs. Most formal training programs also require students to get experience in a commercial kitchen through an internship, apprenticeship, or out-placement program.

Although formal training is an important way to enter the profession, many chefs are trained on the job, receiving real work experience and training from chef-mentors in the restaurants where they work. Others enter the profession through formal apprenticeship programs sponsored by professional culinary institutes, industry associations, and trade unions in coordination with the U.S. Department of Labour. The American Culinary Federation accredits more than 200 formal academic training programs and sponsors apprenticeship programs around the country. Typical apprenticeships last 2 years and combine classroom training and work experience. Accreditation is an indication that a culinary program meets recognized standards regarding course content, facilities, and quality of instruction.

Other qualifications. Chefs, head cooks, and food preparation and serving supervisors must demonstrate strong leadership and communication skills and have the ability to motivate others. Chefs and head cooks also must have an intense desire to cook, be creative, and have a keen sense of taste and smell. Personal cleanliness is essential because most States require health certificates indicating that workers are free from communicable diseases. Knowledge of a foreign language can be an asset because it may improve communication with other restaurant staff, vendors, and the restaurant's clientele.

Certification and advancement. The American Culinary Federation certifies pastry professionals, personal chefs, and culinary educators in addition to various levels of chefs. Certification standards are based primarily on experience and formal training. Although certification is not required, it can help to prove accomplishment and lead to advancement and higher-paying positions. Advancement opportunities for chefs, head cooks, and food preparation and serving supervisors depend on their training, work experience, ability to perform more responsible and sophisticated tasks, and their leadership abilities.

Food preparation and serving supervisors may advance to become food service managers while some chefs and head cooks may go into business as caterers or personal chefs or open their own restaurant. Others may become instructors in culinary training programs, consultants on kitchen design, or food product or equipment sales representatives. A number of chefs and head cooks advance to executive chef positions or food service management positions. When staying in the restaurant business, advancement usually involves moving to a better, busier, or bigger restaurant or working at the corporate level overseeing several restaurants or food service facilities or testing new recipe, menu, or design concepts.

Employment

Chefs, head cooks, and food preparation and serving supervisors held 941,600 jobs in 2008. Food preparation and serving supervisors held 88 percent of these jobs and chefs and head cooks held the remaining 12 percent. Nearly half of chefs and head cooks were employed at full-service restaurants. About nine percent each were employed by hotels and the special food services industry that includes caterers and food service contractors. Eight percent were self-employed.

Forty-three percent of food preparation and serving supervisors were employed by limited-service eating places, made up mostly of cafeterias and fast food restaurants and other places that offer simple carry-out food items. Another 25 percent were employed by full-service restaurants. Supervisors are also found in schools, the special food services industry, and a wide variety of other places that serve food.

Job Outlook

Job opportunities are expected to be good, despite slower than average employment growth, due to the large numbers of workers who leave the occupation and need to be replaced. However, keencompetition is expected for jobs at upscale restaurants that generally pay more.

Employment change. Employment of chefs, head cooks, and food preparation and serving supervisor is expected to increase by 6 percent over the 2008-18 decade, which is more slowly than the average for all occupations.

Growth will be generated by increases in population, a growing variety of dining venues, and continued demand for convenience. As more people opt for the time-saving ease of letting others do the cooking, the need for workers to oversee food preparation and serving will increase. Also, there is a growing consumer desire for healthier, made-from-scratch meals that chefs and head cooks can better prepare.

Job prospects. Job openings for chefs, head cooks, and food preparation and serving supervisors are expected to be good through 2018; however, competition should be keen for jobs at the more upscale restaurants tat tend to pay more. Workers with a good business sense will have beter job prospects, especially at restaurant chains where attention to costs is very important. Although job growth will create many new positions, the majority of job openings will stem from the need to replace workers who leave the occupation. The fast pace, long hours, and high energy levels required for these jobs often lead to high turnover.

Flight attendants

- Competition for positions is expeted to remain keen because the opportunity for travel attracts more applicants than there are jobs.
- Job duties are learned through formal on-the-job training at a flight training centre.
- A high school diploma is the minimum educational requirement, but airlines prefer applicants with a college degree and with experience in dealing with the public.

Nature of the Work

Major airlines are required by law to provide *flight attendants* for the safety and security of the traveling public. Although the primary job of the flight attendants is to ensure that security and safety regulations are followed, attendants also try to make flights comfortable and enjoyable for passenger.

At least 1 hour before takeoff, attendants are biefed by the captain—the pilot in command—on such things as emergency evacuation procedures, coordination of the crew, the length of the flight, expected weather conditions, and any special issues having to do with passengers. Flight attendants make sure that first-aid kits and other emergency equipment are aboard and in working order and that the passenger cabin is in order, with adequate supplesof food, beverages, and any other amenities. As passengers board the plane, flight attendants greet them, check their tickets, and tell them where to store carry-on items.

Before the plane takes off, flight attendants instructall passengers in the use of emergency equipment and check to see that seatbelts are fastened, seat backs are in upright positions, and all carry-on items are properly stowed. In the air, helping passengers in the event of an emergency is the most important responsibility of a flight attendant. Safety-related actions range from reassuring passengers during rough weather to directing passengers who must evacuate a plane following an emergency landing.

Flight attendants also answer questions about the flight, and help small children, elderly or disabled persons, and any others needing assistance. Flight attendants may administer first aid to passengers who become ill. Flight attendants generally serve beverages and on many flights sell precooked meals or snacks. Prior to landing, flight attendants take inventory of headsets, alcoholic beverages, and moneys collected. They also report any medical problems passengers may have had, the condition of cabin equipment, and any lost-and-found articles.

Lead, or first, flight attendants, sometimes known as *pursers*, oversee the work of the other attendants aboard the aircraft, while performing most of the same duties.

Work environment. Because airlines operate around the clock and year round, flight attendants can work nights, holidays, and weekends. In most cases, agreements between the airline and the employees' union determine the total daily and monthly working time. Scheduled on-duty time usually is limited to 12 hours per day, however flight attendants can be scheduled up to 14 hours per day, with somewhat greater maximums for international flying. The Federal Aviation Administration (FAA) requires that flight attendants receive 9 consecutive hours of rest following any duty period.

Attendants usually fly 65 to 90 hours a month and generally spend another 50 hours a month on the ground preparing planes for flights, writing reports following completed flights, and waiting for planes to arrive. Most airlines guarantee a minimum of 65 to 85 flight hours per month, with the option to work additional hours. Flight attendants receive extra compensation for additional hours.

Flight attendants may be away from their home base at least one-third of the time. During this period, the airlines provide hotel accommodations and an allowance for meal expenses.

Flight attendants must be flexible and willing to relocate. However, many flight attendants elect to live in one place and commute to their assigned home base. Home bases and routes worked are bid for and awarded on a seniority basis, so the longer the flight attendant has been employed, the more likely he or she is to work on their preferred flights. Almost all flight attendants start out working on reserve status, or on call. Flight attendants on reserve status usually live near their home base, because they are required to be able to report to their home base on short notice. On small corporate airlines, flight attendants often work on an as-needed basis and must adapt to varying environments and passengers.

The combination of free time and free or discounted airfares provides flight attendants the opportunity to travel. However, the work can be strenuous and trying. Flight attendants stand during much of the flight and must remain pleasant and efficient, regardless of how tired they are or how demanding passengers may be. Occasionally, flight attendants must deal with turbulent flights which can cause difficulties regarding service and cause anxiety among passengers that flight attendants must address.

Working in a moving aircraft leaves flight attendants susceptible to injuries. According to BLS data, full-time flight attendants experienced a much higher than average work-related injury and illness rate. Various physical injuries can occur when opening overhead compartments or while pushing heavy service carts. In addition, medical problems can arise from irregular sleeping and eating patterns, dealing with stressful passengers, working in a pressurized environment, and breathing recycled air.

Training, Other Qualifications, and Advancement

Flight attendants must be certified by the FAA. A high school diploma or its equivalent is the minimum educational requirement, but airlines increasingly prefer applicants who have a college degree. Experience in dealing with the public is important, because flight attendants must be able to interact comfortably with strangers and remain calm under duress.

Education and training. A high school diploma or its equivalent is the minimum educational requirement. However, airlines increasingly prefer applicants with a college degree. Applicants who attend schools or colleges that offer flight attendant training may have an advantage over other applicants. Highly desirable areas of concentration include people-oriented disciplines, such as communications, psychology, nursing, travel and tourism, hospitality, and education. Flight attendants for international airlines generally must speak a foreign language fluently. For their international flights, some of the major airlines prefer candidates who can speak two major foreign languages.

Once hired, all candidates must undergo a period of formal training. The length of training, ranging from 3 to 6 weeks, depends on the size and type of carrier and takes place at the airline's flight training centre. Airlines that do not operate training centres generally send new employees to the centre of another airline. Some airlines may provide transportation to the training centres and an allowance for room, board, and school supplies, while other airlines charge individuals for training. New trainees are not considered employees of the airline until they successfully complete the training program. Trainees learn emergency procedures, such as evacuating an airplane, operating emergency systems and equipment, administering first aid, and surviving in the water. In addition, trainees are taught how to deal with disruptive passengers and with hijacking and terrorist situations.

New hires learn flight regulations and duties, gain knowledge of company operations and policies, and receive instruction on personal grooming and weight control. Trainees for international routes get additional instruction in passport and customs regulations. Trainees must perform many drills and duties unaided, in front of the training staff. Throughout training, they also take tests designed to eliminate unsuccessful trainees. Toward the end of their training, students go on practice flights. Upon successful completion of training, flight attendants receive the FAA Certificate of Demonstrated Proficiency. Flight attendants also are required to go through periodic retraining and pass an FAA safety examination to continue flying.

Licensure and certification. All flight attendants must be certified by the FAA. To be certified, flight attendants are required to successfully complete training requirements, such as evacuation, fire fighting, medical emergency, and security procedures established by the FAA and the Transportation Security Administration. They also must perform the assigned duties of a cabin crew member and complete an approved proficiency check. Flight attendants are certified for specific types of aircraft, regardless of the carrier. Therefore, only 1-day or 2-day recurrent training, with the new carrier, is needed for those flight attendants who change airlines, as long as the type of aircraft remains the same.

Other qualifications. Airlines prefer to hire poised, tactful, and resourceful people who can speak clearly and interact comfortably with strangers and remain calm under duress. Applicants with previous experience in dealing with the public are preferred by airlines. Additionally, airlines usually have age, physical, and appearance requirements. Applicants usually must be at least 18 to 21 years old, although some carriers may have higher minimum-age requirements. Applicants must meet height requirements for reaching overhead bins, which often contain emergency equipment, and most airlines want candidates with weight proportionate to height.

Flight attendants must be in excellent health, and a medical evaluation is required. Vision is required to be correctable to 20/30 or better with glasses or contact lenses (uncorrected no worse than 20/200). Men must have their hair cut above the collar and be clean shaven. Airlines prefer applicants with no visible tattoos, body piercing, or unusual hairstyles or makeup.

In addition to education and training, airlines conduct a thorough background check, which goes back as many as 10 years, as required by the FAA,. Everything about an applicant is investigated, including date of birth, employment history, criminal record, school records, and any gaps in employment. Employment is contingent on a successful background check. An applicant will not be offered a job or will be immediately dismissed if his or her background check shows any discrepancies. All U.S. airlines require that applicants be citizens of the United States or registered aliens with legal rights to obtain employment in the United States.

Advancement. After completing initial training, flight attendants are assigned to one of their airline's bases. New flight attendants are placed on reserve status and are called either to staff extra flights or to fill in for crewmembers that are sick, on vacation, or rerouted. When they are not on duty, reserve flight attendants must be available to report for flights on short notice. They usually remain on reserve for at least 1 year but, in some cities, it may take 5 to 10 years—or longer—to advance from reserve status.

Flight attendants who no longer are on reserve bid monthly for regular assignments. Because assignments are based on seniority, usually only the most experienced attendants get their choice of assignments. Advancement takes longer today than in the past, because experienced flight attendants are remaining in this career longer than in the past.

Some flight attendants become supervisors, moving from senior or lead flight attendant, to check flight attendant, to flight attendant supervisor, then on to base manager, and finally to manager or vice president of in-flight operations. They may take on additional duties, such as recruiting, instructing, or developing in-flight products. Their experience also may qualify them for numerous airline-related jobs involving contact with the public, such as reservation ticket agent or public relations specialist. Flight attendants who do not want to travel often for various reasons may move to a position as an administrative assistant. With additional education, some flight attendants may decide to transfer to other areas of the airline for which they work, such as risk management or human resources.

Employment

Flight attendants held about 98,700 jobs in 2008. Commercial airlines employed the vast majority of flight attendants, and most attendants lived near major metropolitan airports or airports operating as hubs for the major airlines. A small number of flight attendants worked for companies that offered chartered flights.

Job Outlook

Employment of flight attendants is projected to grow about as fast as average. Competition for jobs is expected to remain keen because the opportunity for travel attracts more applicants than there are jobs.

Employment change. Employment of flight attendants is expected to grow by 8 percent, which is about as fast as the average for all occupations over the 2008–18 period. Population growth and an improving economy are expected to boost the number of airline passengers. As airlines expand their capacity to meet rising demand by increasing the number and size of planes in operation and the number of flights offered, more flight attendants will be needed.

Job prospects. Despite growing demand for flight attendants, competition is expected to be keen because this job usually attracts more applicants than there are jobs, with only the most qualified eventually being hired. College graduates who have experience dealing with the public should have the best chance of being hired. Job opportunities may be better with the faster growing regional and commuter, low-cost, and charter airlines. There also are job opportunities for professionally trained flight attendants to work for companies operating private aircraft for their executives.

The majority of job opportunities through the year 2018 will arise from the need to replace flight attendants who leave the labour force or transfer to other occupations, often for higher earnings or a more stable lifestyle. With the job now viewed increasingly as a profession, however, fewer flight attendants leave their jobs, and job turnover is not as high as in the past. According to the Association of Flight Attendants, the average job tenure of attendants is currently 16 years and is increasing.

In the long run, opportunities for persons seeking flight attendant jobs should improve as the airline industry expands. Over the next decade, however, demand for flight attendants will fluctuate with the demand for air travel, which is highly sensitive to swings in the economy. During downturns, as air traffic declines, the hiring of flight attendants declines, and some experienced attendants may be laid off until traffic recovers.

Earnings

Median annual wages of flight attendants were $35,930 in May 2008. The middle 50 percent earned between $28,420 and $49,910. The lowest 10 percent earned less than $20,580, and the highest 10 percent earned more than $65,350. According to data from the Association of Flight Attendants, beginning attendants had median earnings of $16,191 a year in 2009. Beginning pay scales for flight attendants vary by carrier, however. New hires usually begin at the same pay scale, regardless of experience; all flight attendants receive the same future pay increases based on an established pay scale.

Some airlines offer incentive pay for working holidays, night and international flights, or taking positions that require additional responsibility or paperwork.

Flight attendants and their immediate families are entitled to free or discounted fares on their own airline and reduced fares on most other airlines. Some airlines require that the flight attendant be with an airline for 3 to 6 months before taking advantage of this benefit. Other benefits may include medical, dental, and life insurance; 401K or other retirement plan; sick leave; paid holidays; stock options; paid vacations; and tuition reimbursement. Flight attendants also receive a "per diem" allowance for meal expenses while on duty away from home. Flight attendants are required to purchase uniforms and wear them while on duty. The airlines usually pay for uniform replacement items, and may provide a small allowance to cover cleaning and upkeep of the uniforms.

The majority of flight attendants hold union membership, primarily with the Association of Flight Attendants. Other unions that represent flight attendants include the Transport Workers Union of America and the International Brotherhood of Teamsters.

Retail Salespersons

- Good employment opportunities are expected because of the need to replace the large number of workers who leave the occupation each year.
- Many salespersons work evenings and weekends, particularly during peak retail periods.
- Employers look for people who enjoy working with others and who have good communication skills, an interest in sales work, a neat appearance, and a courteous demeanor.
- Although advancement opportunities are limited, having a college degree or a great deal of experience may help retail salespersons move into management positions.

Nature of the Work

Whether selling shoes, computer equipment, or automobiles, retail salespersons assist customers in finding what they are looking for. They also try to increase sales by describing a product's features, demonstrating its uses, and promoting its value.

In addition to selling, many retail salespersons—especially those who work in department and apparel stores—conduct financial transactions with their customers. This usually involves receiving payments by cash, check, debit card, or credit card; operating cash registers; and bagging or packaging purchases. Depending on the hours they work, retail salespersons may have to open or close cash registers. This work may include counting the money in the register and separating charge slips, coupons, and exchange vouchers.

Retail salespersons also may have to make deposits at a cash office. (Cashiers, who have similar duties, are discussed elsewhere in the *Handbook*.) In addition, retail salespersons may help stock shelves or racks, arrange for mailing or delivery of purchases, mark price tags, take inventory, and prepare displays.

For some sales jobs, particularly those involving expensive and complex items, retail salespersons need special knowledge or skills.

For example, salespersons who sell automobiles must be able to explain the features of various models, the manufacturers' specifications, the types of options and financing available, and the details of associated warranties. In addition, all retail salespersons must recognize security risks and thefts and understand their organization's procedure for handling such situations—procedures that may include notifying security guards or calling police. Work environment. Most retail salespersons work in clean, comfortable, well-lit stores. However, they often stand for long periods and may need supervisory approval to leave the sales floor. They also may work outdoors if they sell items such as cars, plants, or lumber yard materials.

The Monday-through-Friday, 9-to-5 workweek is the exception rather than the rule for retail salespersons. Many salespersons work evenings and weekends, particularly during holidays and other peak sales periods. The end-of-year holiday season often is the busiest time, and as a result, many employers limit the use of vacation time between Thanksgiving and the beginning of January.

This occupation offers opportunities for both full-time and part-time work. About 34 percent of retail salespersons worked part time in 2008. Part-time opportunities may vary by setting, however, as many who sell big-ticket items are required to work full time.

Training, Other Qualifications, and Advancement

Retail salespersons typically learn their skills through on-the-job training. Although advancement opportunities are limited, having a college degree or a great deal of experience may help retail salespersons move into management positions. Education and training. There usually are no formal education requirements for retail sales positions, but employers often prefer applicants with a high school diploma or its equivalent. This may be especially important for those who sell technical products or "big-ticket" items, such as electronics or automobiles. A college degree may be required for management trainee positions, especially in larger retail establishments.

Most retail salespersons receive on-the-job training, which usually lasts anywhere from a few days to a few months. In small stores, newly hired workers usually are trained by an experienced employee. In large stores, training programs are more formal and generally are conducted over several days. Topics often include customer service, security, the store's policies and procedures, and cash register operation. Depending on the type of product they are selling, employees may be given additional specialized training. For example, those working in cosmetics receive instruction on the types of products the store offers and for whom the cosmetics would be most beneficial. Likewise, those who sell computers may be instructed in the technical differences between computer products. Because providing the best possible service to customers is a high priority for many employers, employees often are given periodic training to update and refine their skills.

Other qualifications. Employers look for people who enjoy working with others and who possess good communication skills. Employers also value workers who have the tact and patience to deal with difficult customers. Among other desirable characteristics are an interest in sales work, a neat appearance, and a courteous demeanor. The ability to speak more than one language may be helpful for employment in communities where people from various cultures live and shop. Before hiring a salesperson, some employers conduct a background check, especially for a job selling high-priced items.

Advancement. Opportunities for advancement vary. In some small establishments, advancement is limited because one person—often the owner—does most of the managerial work. In others, some salespersons can be promoted to assistant manager. Large retail businesses usually prefer to hire college graduates as management trainees, making a college education increasingly important. However, motivated and capable employees without college degrees still may advance to administrative or supervisory positions in large establishments.

As salespersons gain experience and seniority, they often move into positions with greater responsibility and may be given their choice of departments in which to work. This opportunity often means moving to areas with higher potential earnings and commissions. The highest earnings potential usually lies in selling "big-ticket" items—such as cars, jewellery, furniture, and electronic equipment—although doing so often requires extensive knowledge of the product and an excellent talent for persuasion.

Previous sales experience may be an asset when one is applying for positions with larger retailers or in nonretail industries, such as financial services, wholesale trade, or manufacturing.

Employment

Retail salespersons held about 4.5 million jobs in 2008. The largest employers were clothing and clothing accessories stores, department stores, building material and supplies dealers, motor vehicle and parts dealers, and general merchandise stores such as warehouse clubs and supercenters. In addition, about 156,500 retail salespersons were self-employed. Because retail stores are found in every city and town, employment is distributed geographically in much the same way as the population.

Job Outlook

Employment is expected to grow about as fast as average. Due to the frequency with which people leave this occupation, job opportunities are expected to be good. Employment change. Employment is expected to grow by 8 percent over the 2008–18 decade, about as fast as the average for all occupations. In addition, given the size of this occupation, about 374,700 new retail salesperson jobs will arise over the projections decade—more jobs than will be generated in almost any other occupation.

Employment growth among retail salespersons reflects rising retail sales stemming from a growing population. Many retail establishments will continue to expand in size and number, leading to new retail sales positions. Growth will be fastest in general merchandise stores, many of which sell a wide assortment of goods at low prices. As consumers continue to prefer these stores other establishments with higher prices, growth in this industry will be rapid. Employment of retail sales persons is expected to decline in department stores and automobile dealers as these industries see a reduction in store locations.

Despite the growing popularity of electronic commerce, the impact of online shopping on the employment of retail salespersons is expected to be minimal. Internet sales have not decreased the need for retail salespersons. Retail stores commonly use an online presence to complement their in-store sales, and many consumers prefer to buy merchandise in person. Retail salespersons will remain important in assisting customers, providing specialized service, and increasing customer satisfaction.

Job prospects. Employment opportunities for retail salespersons are expected to be good because of the need to replace the large number of workers who transfer to other occupations or leave the labour force each year. In addition, many new jobs will be created for retail salespersons as businesses seek to expand operations and enhance customer service. A substantial number of these openings should occur in warehouse clubs and supercenters as a result of strong growth among these establishments.

Opportunities for part-time work should be abundant, and demand is expected be strong for temporary workers during peak selling periods, such as the end-of-year holiday season between Thanksgiving and the beginning of January. During economic downturns, sales volumes and the resulting demand for sales workers usually decline.

Consequently, retail sales jobs generally are more susceptible to fluctuations in the economy than are many other occupations.

Earnings

Median hourly wages of wage-and-salary retail salespersons, including commissions, were $9.86 in May 2008. The middle 50 percent earned between $8.26 and $13.35 an hour. The lowest 10 percent earned less than $7.37, and the highest 10 percent earned more than $19.14 an hour. Median hourly wages in the industries employing the largest numbers of retail salespersons in May 2008 were as follows:

Automobile dealers	$18.91
Building material and supplies dealers	11.95
Other general merchandise stores	9.22
Department stores	9.14
Clothing stores	8.94

Many beginning or inexperienced workers earn the Federal minimum wage of $7.25 an hour, but many States set minimum wages higher than the Federal minimum. In areas where employers have difficulty attracting and retaining workers, wages tend to be higher than the legislated minimum.

Compensation systems can vary by type of establishment and merchandise sold. Salespersons receive hourly wages, commissions, or a combination of the two. Under a commission system, salespersons receive a percentage of the sales they make. This system offers sales workers the opportunity to increase their earnings considerably, but they may find that their earnings depend strongly on their ability to sell their product and on the ups and downs of the economy.

Benefits may be limited in smaller stores, but benefits in large establishments usually are considerable. In addition, nearly all salespersons are able to buy their store's merchandise at a discount, with the savings depending on the type of merchandise. Also, to bolster revenue, employers may use incentive programs such as awards, bonuses, and profit-sharing plans to the sales staff.

Housekeeping

Housekeeping or housecleaning is the systematic process of making a home neat and clean in approximately that order. This maybe applied more broadly that just an individual home, or as a metaphor for a similar "clean up" process applied elsewhere such as a procedural reform. It can also be called household management, which is the act of overseeing the organizational, financial, day-to-day operations of a house or estate, and the managing of other domestic concerns.

In the process of housekeeping general cleaning activities are completed, such as disposing of rubbish, storing of belongings in regular places, cleaning dirty surfaces, dusting and vacuuming. It is also the care and control of property, ensuring its maintenance and proper use and appearance. In a hotel, "housekeeping" is also a term for the cleaning personnel.

Some housekeeping is housecleaning and some housekeeping is home chores. Home chores are housework that needs to be done at regular intervals, Housekeeping includes the budget and control of expenditures, preparing meals and buying food, paying the heat bill, and cleaning the house. Outdoor housecleaning chores include removing leaves from rain gutters, washing windows, sweeping doormats, cleaning the pool, putting away lawn furniture, and taking out the trash.

Tools include the vacuum cleaner, broom and mop. Supplies such as cleaning solutions and sponges are sold in grocery stores and elsewhere. Professional cleaners can be hired for less frequent or specialist tasks such as cleaning blinds, rugs, and sofas. Professional services are also offered for the basic tasks. Safety is a consideration because some cleaning products are toxic and some cleaning tasks are physically demanding. *Green cleaning* refers to cleaning without causing pollution. The history of housecleaning has links to the advancement of technology.

Reasons

People perform house cleaning for the home to look better and be safer and easier to live in. It is in response to clutter, disorder, litter, dirtiness or to prevent such. Without housecleaning limescale builds up on taps, mold grows in wet areas, bacterial action make the garbage disposal and toilet smell and cobwebs accumulate. With organisation belongings are easily found and table tops are clear.

Indoor Litter

Disposal of rubbish is an important aspect of house cleaning, the reasons for this are psychological, social and practical. Plastic bags are designed and manufactured specifically for the collection of litter. Many are sized to fit common waste baskets and trash cans. Paper bags are made to carry aluminum cans, glass jars and other things. Recycling is possible with some kinds of litter.

Clutter Problem

Clutter is belongings that have not been put away into storage locations designed for them. If there are not enough shelves and drawers and hangers and there can't be more, this is a cause of clutter. There is a limit to the number of possessions that can be neatly stored in a home. A tangled pile of old coats, mittens, scarves, hats and boots occurs in some hall closets. Pawnbrokers, thrift shops and garbage collectors are involved in the prevention of clutter.

Dustiness

Over time dust accumulates on household surfaces. As well as making the surfaces dirty, when dust is disturbed it can become suspended in the air, causing sneezing and breathing trouble. It can also transfer from furniture to clothing, making it unclean. Various tools have been invented for dust removal; Feather and lamb's wool dusters, cotton and polyester dust cloths, disposable paper "dust cloths", dust mops for smooth floors and vacuum cleaners. Vacuum cleaners often have a variety of tools to enable them to remove not just from carpets and rugs, but from hard surfaces and upholstery. This way can help your things well organize. Things will look so clean likewise neat.

Dirtiness

Although one meaning of housecleaning is "improvement," the primary meaning of housecleaning is "the cleaning of a house." A goal of housecleaning would be the removal of mold from shower grout or smudges and splatters from a kitchen wall. Examples of dirtiness or "soil" would be dry coffee spills and jelly drips or muddy footprints on carpet. Soap and water is a "cleaner." Equipment used with a cleaner might be a bucket and sponge. A modern tool is the spray bottle, but the scientific principle is the same. Cleaning supplies have directions and are sold at janitorial and other stores. Cleaning specialists such as carpet cleaners are listed in phone books. Housecleaning is done to achieve and to enjoy a cleaner house.

Training

In the early 1800s throughout the industrialized world there was a campaign to teach girls the domestic arts (cooking, cleaning, hygiene, sewing, art, decor, etc) in school. In the early 1870s there were college courses in home economics and by 1880 there were high school courses. There continues to be high school and vocational school courses and college degrees in home economics, which prepares students for various employment, as well as home and family management. Home economists are taught and teach about relationships, children, economy, shopping, management of home and family, sewing and interior decoration. Although boys have not been required to learn cooking and cleaning in school, a few have taken a course or two.

Household Chemicals

Various household cleaning products have been developed to facilitate the removal of dust and dirt, for surface maintenance, and for disinfection. Products are available in powder, liquid or spray form. The basic ingredients determine the type of cleaning tasks for which they are suitable. Some are packaged as general purpose cleaning materials whilst others are targeted at specific cleaning tasks such as drain clearing, oven cleaning, lime scale removal and polishing furniture. Household cleaning products provide aesthetic and hygiene benefits but are also associated with health risks for the users, and building occupants. The US Department of Health and Human Services offers the public access to the Household Products Database.

This database provides consumer information for over 4,000 products based on information provided by the manufacturer through the Material Safety Data Sheet. Consumers can search for products by brand name, manufacturer, ingredients, product recalls, and health effects. Surfactants lower the surface tension of water, making it able to flow into smaller tiny cracks and crevices in soils making removal easier. Alkaline chemicals break down known soils such as grease and mud. Acids break down soils such as lime scale, soap scum, and stains of mustard, coffee, tea, and alcoholic beverages. Some solvent-based products are flammable and some can dissolve paint and varnish. Disinfectants stop smell and stains caused by bacteria.

When multiple chemicals are applied to the same surface without full removal of the earlier substance, the chemicals may interact. This interaction may result in a reduction of the efficiency of the chemicals applied (such as a change in pH value caused by mixing alkalis and acids) and in cases may even emit toxic fumes. An example of this is the mixing of ammonia-based cleaners (or acid-based cleaners) and bleach. This causes the production of chloramines that volatilize (become gaseous) causing acute inflammation of the lungs (toxic pneumonitis), long-term respiratory damage, and potential death.

Residue from cleaning products and cleaning activity (dusting, vacuuming, sweeping) have been shown to impact indoor air quality (IAQ) by redistributing particulate matter (dust, dirt, human skin cells, organic matter, animal dander, particles from combustion, fibers from insulation, pollen, and polycyclic aromatic hydrocarbons that gaseous or liquid particles become adsorbed to. The particulate matter and chemical residual will of be highest concentrations right after cleaning but will decrease over time depending upon levels of contaminants, air exchange rate, and other sources of chemical residual. Of most concern are the family of chemicals called VOCs such as formaldehyde, toluene, and limonene.

Volatile organic compounds (VOCs) are released from many household cleaning products such as disinfectants, polishes, floor waxes, air-freshening sprays, all purpose cleaning sprays, and glass cleaner. These products have been shown to emit irritating vapors. VOCs are of most concern due to their tendency to evaporate and be inhaled into the lungs or adsorbed to existing dust, which can also be inhaled. It has been found that aerosolized (spray) cleaning products are important risk factors and may aggravate symptoms of adult asthma, respiratory irritation, childhood asthma, wheeze, bronchitis, and allergy.

Other modes of exposure to potentially harmful household cleaning chemicals include absorption through the skin (dermis), accidental ingestion, and accidental splashing into the eyes. Products for the application and safe use of the chemicals are also available, such as nylon scrub sponge and rubber gloves. It is up to the consumer to keep themselves safe while using these chemicals. Reading and comprehending the labels is important.

There is a growing consumer and governmental interest in natural cleaning products and green cleaning methods. The use of nontoxic household chemicals is growing as consumers become more informed of the health effects of many household chemicals, and municipalities are having to deal with the expensive disposal of household hazardous waste (HHW).

Tools

"Modern housecleaning tools" is almost an oxymoron. There are few areas of employment where someone from 50 years ago could step into the same job today, but housecleaning is one area where there has been very little change. Brooms remove debris from floors. Brushes clean cracks and crevices. Buckets hold cleaning and rinsing solutions. Carpet sweepers remove surface dust and debris. Chamois do lint-free drying of windows. Clotheslines hold doormats and rugs for cleaning. Clothespins fasten things on clotheslines. Dishwashers are machines that wash dishes placed inside them. Dryers dry textiles placed inside them. Dusters are soft cloths for wiping dust, or are various dust sweepers. Dustpans carry dust and debris swept into them.

Floor machines remove and apply floor finish. Rubber gloves protect hands from dish water and other cleaners. Dust mops remove dust from smooth floors. Wet mops are for washing floors. Paper towels wipe up grease and other materials difficult to rinse from a cloth. Polishing cloths are for applying polish and removing excess polish. Sponges apply cleaning solution and remove soil. Spray bottles apply cleaning solution to be removed by another tool. Squeegees remove solutions from glass. Steam cleaners are machines that help wash carpet. Vacuum cleaners remove dust from inside carpet. *Upright vacuums* are best at carpet vacuuming. *Canister vacuums* are best at upholstery and bare floors. *Wet-dry vacuums* can be used to remove spills and spot removers. Washing machines clean textiles such as dust cloths.

Yard

A home's yard and exterior are sometimes subject to cleaning. The first impression of a home's cleanliness is given by its yard and entranceway. Exterior cleaning also occurs for safety, upkeep and usefulness. It includes removal of paper litter and grass growing in sidewalk cracks. Rain gutters, doormats, pools and the screens and glass of windows are also cleanable. Yard junk-removal might occur and porch clutter removal. The paint of door frames might be washed or an old pinata thrown away.

Professional Cleaning Services

There are many professional house cleaning services located around the world. In recent years, they have grown in popularity due to the increase in both spouses needing to work.

Investment in Property

Funding Real Estate Projects for the Hospitality Industry: Emerging Perspectives

Up until 2007, most hoteliers, investors and developers were buoyant when it came to the growth prospects in the hospitality industry. They had enough reason to be optimistic as every factor which would influence the industry, directly or indirectly, was on a growth trajectory. The GDP was growing like never before and the whole world had its eyes on the Indian growth story. Plans of expansion filled the newspapers and press releases and investors were keener than ever to get a fair share of the pie.

The recession, however, had plans of its own and devoured most of the pie. The global economic slowdown and its effect on the Indian economy has doused the fire of excitement of even the most optimistic developers and investors. It has resulted in an extreme crunch for investment in the hospitality sector, coupled with the decrease in demand for rooms. This double whammy put to rest most of the ambitious plans of expansion across the country. All players have been reviewing their plans of development, owing to the increasingly challenging macro economic situation at the moment. The total number of rooms estimated to be added is today nearly half of what was announced earlier. One-fourth of the announced plans have fallen completely flat and the rest are hanging on the edge of viability. DLF, Parsvnath and other developers of similar cadre have scaled down or slowed down their plans of expansion.

Parsvnath, which had plans of adding at least 10,000 rooms, has now stopped acquiring land for any further plans other than the twenty hotels for which they have already done the same. There have been reports that DLF has been in talks with various hotel companies to sell eight to nine of their land parcels demarcated for hotel projects to raise funds. Unitech has sold its Gurgaon hotel project to reduce its huge debt burden. Developers are keener to finish the projects on hand rather than plan further. Divesting the investment heavy hotel plans seems to be the best way out for the cash strapped, heavily indebted players to survive the present-day economic scenario.

Financial Projections Going Awry

The seeds were sown, the crops were nurtured through the tough inflationary times but when the time to harvest came, the floods of recession washed away the anticipated bounty. Cost and revenue assumptions made during the good times have thus gone for a toss. When it comes to loan disbursements, real estate is presently the black sheep of the family.

Private banks from which loans were freely available earlier have dried up. Public sector banks which continue to lend, albeit cautiously, now require a higher collateral to lend the same amount.

Non-banking finance companies are either not lending at all or looking at returns in the post 20 per cent range. Private equity interest in the hospitality sector has all but dried up. Due to the severe global liquidity crunch and flight of capital to 'places of origin', there is a diminishing interest for private equity players in foreign markets. This has added to the financial woes of the capital thirsty developers.

The risk associated with a hospitality project being relatively larger, the premium at which funding is available has gone up. Due to this, only the most stable projects in the market would be able to take up the risk of delivering higher returns to the lenders. Many developers had invested heavily in land in the past when the land values were significantly higher. At present, the value of the same land parcels has come down significantly. The dependency on the appreciation of these assets has turned out to be a major dampener to the development plans of the various developers. The lower value of land means that the value of collateral has come down for project loans.

The cycle of cost of construction has taken the industry on a roller coaster ride over the past couple of years. The global commodity cycle has drastically changed its course. Steel and cement which form a significant chunk of the civil construction costs have lost up to 40 per cent from their historic highs twelve months ago.

The hotels which have opened recently have faced the brunt of the cost fluctuations in a similar fashion. Greater costs were incurred, owing to the period in which their construction phases passed through. The cost of materials was higher, the market was booming and along with the high material costs, the various architects and consultants demanded a premium as a result of a never-before-seen demand. Despite the increased amount of investment that the developers had to put in, they now face a world with reduced revenue prospects.

The main factor that has directly and indirectly influenced the stability of revenue-side financial projections in the hotel industry is the lack of 'stickiness' relative to other real estate sectors.

Stickiness of the hotel industry is low. In order to explain, take lease agreements into consideration. These are long term in nature and hence revenues are more secure in the case of an office or retail space as they have a considerable lock-in period. However in case of the hospitality industry, the revenues are more directly susceptible to market conditions. This being the reason as to why other sectors have been relatively less affected by the present day scenario.

The main factors influencing this stickiness are occupancy rate and Average Room Rate (ARR).

Occupancy rate is a function of supply and demand. The present scenario is affected by both the demand and supply factors, with demand having the more potent influence. The two main demand drivers, the leisure and business travellers, have contributed to the reduction in occupancy rates. Overseas travel has been affected to a large extent due to the economic slowdown.

Corporates have found innovative ways to cut down on their expenses and there is a growing need for them to rationalise business travel. They have reduced their travel budgets, are staying at serviced apartments and guesthouses or even looking at options where they can avoid staying overnight. MNCs with significant exposure to the developed markets are taking the lead to drive travel associated business expenditures down as a part of their global strategy. Spending sentiments of consumers has been hit to a large extent owing mostly to fears of job loss and a resulting lack of confidence and low morale. The leisure consumers are hence looking to spend lesser on travel.

In addition to the demand slowing down, the supply is on the increase which would mean that the occupancy rates are set to reduce further at least for the next one or two years. All these factors contribute to the occupancy rate reducing significantly and this is a major reason contributing to the instability of the financial projections made by the various players. On the supply side, although there is an inherent demand for more supply in the long run, occupancy rates would see an improvement only with the revival of the economy.

Due to the hit on occupancy rate, hoteliers have been forced to cut down on the ARRs to attract both their business and leisure customers. The occupancy rate has also negatively affected the other closely linked revenue sources such as food and beverage, conferences and banquets.

Role of Funding Options

Traditionally, most developers have tended to plough back the surplus that they earn, into their business via investing in land banks. This makes them highly dependant on external sources for funding. The various equity funding options available in the market till now have been public or private equity which can be foreign (FDI) or domestic funding. In the present market scenario, there is a lot more uncertainty in cash flows associated with hotel projects than usual. This has led investors to be quite cautious when it comes to investing in or lending to hospitality projects.

With many of the real estate companies trading below book values, the public equity scenario is dismal. IPO market is almost non-existent. The last high profile real estate IPO which failed was the EMAAR MGF IPO in February 2008. Real Estate Index has fallen up to 80 per cent from its peak. Going ahead with secondary offerings or rights issues would likely meet with a negative investor response.

Private equity, while still being an option, has seen a slowdown. Investors are worried about the market bottoming out and there is a feeling that the correction in the hospitality sector is still not complete. PE players are avoiding common equity in SPVs but looking at structured investments with greater security and preferred returns. However, there are a number of funds which are actively looking to buy out or invest in distressed assets at enticing valuations.

Having said that, owing to the severe capital crunch, the market for new PE capital raising especially in real estate is difficult right now and is likely to remain so for the next two to three years. Where funds have already been raised, there have been cases of Limited Partners (LPs) not honouring their capital commitments.

Project loans for under construction projects are harder to come by with only the public sector banks lending. These loans are being disbursed primarily to promising projects with substantial asset cover guarantees. We are also witnessing cases of liquidity parched developers borrowing from HNIs at extremely high rates.

Bailing out Stranded Projects

A million dollar question on everyone's mind would be on how to bail oneself out of stranded projects.

Divesting a part of the stake in the project to gain capital may be one of the options. This may not be very easy in today's market. The project maybe valued at a much lower price than what is expected and may leave the seller with a raw deal from the transaction made. One could also think of repositioning the project. Instead of increasing the investment requirement for a project by planning a high end luxury hotel, one could look at serviced apartments and budget hotels at the moment to get through the times of credit crunch.

Another option would be to reduce the scale of the project. This can be done in one go or in phases. Phasing the project out in stages where part of the hotel could be operational in a relatively lesser time and with lesser investment than earlier planned would help ease the credit and liquidity crunch. There are quite a few operators who take a stake in the project as well. Accor is a good example of such an operator who have significant expansion plans in India. Tying up with the right investing operators would however not be as easy as it may look as these operators would choose only the best of the options available and they would have their own plans in place already.

Going back to the lender may be an option worth visiting, to see if they could restructure the existing loan. One could also go for refinancing or extending the loan. If the borrower has a good record, they could also look at finding a new lender as well, probably an HNI who would be willing to invest. However, if the only option left is to exit the project, the timing would need to be well thought out. If one waits too long for a good price, the price might just go lower. On the other hand one also needs to assess the urgency to exit the project as we would not want to end up in a situation where we would be forced to accept a price which would normally be unattractive.

Liberalisation of FDI Norms and its Impact on Expansion Plans

FDI has the promise to be a major factor in the economic development of the many developing nations of the world. With tourism being a major revenue earner for countries across the world, FDI norms liberalisation in this sector is a definite boost for the Indian economy. The government has been very liberal when it comes to FDI regulations in the hotel and tourism industry. According to the Government of India-Ministry of Commerce and Industry, 100 per cent FDI is permissible in the hotel and tourism sector on the automatic route subject to the automatic approval clauses.

The additional restrictions, applicable to some sectors of real estate, such as area of development being at least 50,000 square metres, 50 per cent of the project to be completed before five years, no repatriation of funds before three years from date of minimum capitalisation released in 2005, are not applicable to the hotel and tourism industry. The government has thus made a conscious effort knowing the fact that tourism can be a major source of revenue for the country which is still relatively untapped. Hence, the only effect that existing FDI norms can have on the industry is on the positive side. It is only a matter of time before investments in the hotel and tourism sector start flowing in freely once again.

These investments would be more dependant on the inherent demand in the industry and the overall economic scenario. Once the credit crunch situation gets better, the inherent demand in the industry that one senses should help attract funds easily as compared to some of the other sectors. Some say the worst is over and India will be getting back on its feet soon. However, the key is not just to wait and watch but to try and be one step ahead. Be ahead and decide on when to stop waiting. Only the best would survive and we would be lying if we said we weren't all eager to see who they really are.

The Hotel Organisation Framework

Both people and organizations need to establish a strategic framework for significant success. This framework consists of:

- a vision for your future,
- a mission that defines what you are doing,
- values that shape your actions,
- strategies that zero in on your key success approaches, and
- goals and action plans to guide your daily, weekly and monthly actions.

Your organization's success and your personal success depend on how well you define and live by each of these important concepts. In fact:

- Companies whose employees understand the mission and goals enjoy a 29 percent greater return than other firms (Watson Wyatt Work Study).
- U.S. workers want their work to make a difference, but 75% do not think their company's mission statement has become the way they do business.

Read more to find out how to develop a successful strategic framework for your organization and yourself.

Vision Statement

A vision is a statement about what your organization wants to become. It should resonate with all members of the organization and help them feel proud, excited, and part of something much bigger than themselves. A vision should stretch the organization's capabilities and image of itself. It gives shape and direction to the organization's future. Visions range in length from a couple of words to several pages. I recommend shorter vision statements because people will tend to remember their shorter organizational vision.

Vision Statement Samples

"Year after year, Westin and its people will be regarded as the best and most sought after hotel and resort management group in

North America." "To be recognized and respected as one of the premier associations of HR Professionals."

Personal Vision Statement

Your personal vision for your life can be as simple as a couple of words or as lengthy as 200 or more items you want to attain or accomplish.

Mission or Purpose is a precise description of what an organization does. It should describe the business the organization is in. It is a definition of "why" the organization exists currently. Each member of an organization should be able to verbally express this mission. Additionally, each person needs a mission for his or her life. The alignment of your life mission with your organization's mission is one of the key factors in whether you are happy with your work and workplace. If they are incongruent, you are likely dissatisfied with your work choice.

Mission Statement Samples

"Our goal is simply stated. We want to be the best service organization in the world." (IBM)

"FedEx is committed to our People-Service-Profit Philosophy. We will produce outstanding financial returns by providing totally reliable, competitively superior, global, air-ground transportation of high-priority goods and documents that require rapid, time-certain delivery."

"To give ordinary folk the chance to buy the same thing as rich people."

"Our mission is to earn the loyalty of Saturn owners and grow our family by developing and marketing U.S.-manufactured vehicles that are world leaders in quality, cost, and customer enthusiasm through the integration of people, technology, and business systems." "In order to realize our Vision, our Mission must be to exceed the expectations of our customers, whom we define as guests, partners, and fellow employees.(mission) We will accomplish this by committing to our shared values and by achieving the highest levels of customer satisfaction, with extraordinary emphasis on the creation of value. (strategy) In this way we will ensure that our profit, quality and growth goals are met." (Westin Hotels and Resorts)

Values are traits or qualities that are considered worthwhile; they represent an individual's highest priorities and deeply held driving forces. (Values are also known as core values and as governing values;

they all refer to the same sentiment.) Value statements are grounded in values and define how people want to behave with each other in the organization. They are statements about how the organization will value customers, suppliers, and the internal community.

Value statements describe actions which are the living enactment of the fundamental values held by most individuals within the organization.

The values of each of the individuals in your workplace, along with their experience, upbringing, and so on, meld together to form your corporate culture. The values of your senior leaders are especially important in the development of your culture.

These leaders have a lot of power in your organization to set the course and environment and they have selected the staff for your workplace. If you think about your own life, your values form the cornerstones for all you do and accomplish. They define where you spend your time, if you are truly living your values. Each of you makes choices in life according to your most important four – ten values. Why not take the time to identify what is most important to you and to your organization.

Why Identify and Establish Values?

Effective organizations identify and develop a clear, concise and shared meaning of values/beliefs, priorities, and direction so that everyone understands and can contribute. Once defined, values impact every aspect of your organization. You must support and nurture this impact or identifying values will have been a wasted exercise.

People will feel fooled and misled unless they see the impact of the exercise within your organization. If you want the values you identify to have an impact, the following must occur.

- People demonstrate and model their values in action in their personal work behaviours, decision making, contribution, and interpersonal interaction.
- Organizational values help each person establish priorities in their daily work life.
- Values guide every decision that is made once the organization has cooperatively created the values and the value statements.
- Rewards and recognition within the organization are structured to recognize those people whose work embodies the values the organization embraced.

- Organizational goals are grounded in the identified values. Adoption of the values and the behaviours that result is recognized in regular performance feedback.
- People hire and promote individuals whose outlook and actions are congruent with the values.
- Only the active participation of all members of the organization will ensure a truly organization-wide, value-based, shared culture.

The following are examples of values: ambition, competency, individuality, equality, integrity, service, responsibility, accuracy, respect, dedication, diversity, improvement, enjoyment/fun, loyalty, credibility, honesty, innovativeness, teamwork, excellence, accountability, empowerment, quality, efficiency, dignity, collaboration, stewardship, empathy, accomplishment, courage, wisdom, independence, security, challenge, influence, learning, compassion, friendliness, discipline/order, generosity, persistence, optimism, dependability, flexibility.

Although important aspects of your life and attention, these are not values: family, church, professionalism. If you define what you value about each of these, you are identifying the core value. As an example, the core value in family might be close relationships; in church, spirituality.

Want to see samples of values and value statements? These give you an idea of the depth and breadth in which organizations write their values.

Search online for values and value statements and you'll find some that stretch to several pages, too.

At Merck, "corporate conduct is inseparable from the conduct of individual employees in the performance of their work. Every Merck employee is responsible for adhering to business practices that are in accordance with the letter and spirit of the applicable laws and with ethical principles that reflect the highest standards of corporate and individual behavior...

"At Merck, we are committed to the highest standards of ethics and integrity. We are responsible to our customers, to Merck employees and their families, to the environments we inhabit, and to the societies we serve worldwide.

In discharging our responsibilities, we do not take professional or ethical shortcuts. Our interactions with all segments of society must reflect the high standards we profess.".

Our Employees

- Our employees are the most valued assets of our company, essential participants with a shared responsibility in fulfilling our mission.
- We recognize that the quality, motivation and performance of our employees are the key factors in achieving our success.

  "Accordingly, our Human Resources policies and practices are built on:
- Dedication to assisting every employee in reaching his or her full potential in both performance and reward.
- Commitment to diversity, equal opportunity and fair treatment.
- Promotion based on merit, and from within whenever possible.
- We want our organizational structure and culture to promote employee involvement, open communication, teamwork and cooperation."

Strategies are the broadly defined four or five key approaches the organization will use to accomplish its mission and drive toward the vision.

Goals and action plans usually flow from each strategy. One example of a strategy is employee empowerment and teams. Another is to pursue a new worldwide market in Asia. Another is to streamline your current distribution system using lean management principles. A university Human Resources Development department established several broad strategies for growth. These included becoming the training and education resource of choice for all employees by offering one-stop access to any and all existing education and training resources. Additionally, they determined key strategies for expanding their funding base and moving courses online for customer convenience.

A Human Resources department devised strategies to develop a superior workforce. These included eliminating poor performers; hiring from several choices of excellent candidates, not just "settling" on a candidate; developing succession planning; and increasing training and cross-training opportunities.

Sample Strategies

"The Human Resource Association of Greater Detroit's (HRAGD) efforts to advance its mission will include: The promotion of voluntary member interchange, observance of ethical and professional standards, the conduct of meetings and workshops on relevant human resources topics and issues, communication of our purpose and activities to the broader business community, cooperation with the Society for Human Resources Management (SHRM), as well as, other SHRM professional and student chapters and related Human Resources organizations and the community involvement of our membership.

The Association regularly publishes newsletters throughout the year which cover items such as monthly meeting highlights, future programs, Executive Board announcements, SHRM and legislative updates and general human resources news. In addition, a Membership Directory and member skills listing are published."

The San Antonio Express News developed these strategies.

- "Expand our customer base and enhance the franchise by pursuing multimedia opportunities.
- Deliver an award-winning level of journalistic excellence, building public interest, trust and pride.
- Provide vigorous community leadership and support.
- Instill an environment of internal and external excellence in customer service.
- Empower and recognize each employee's unique contribution.
- Achieve the highest standards of quality.
- Improve financial strength and profitability."

Goals and Action Plans

After you have developed the key strategies, turn your attention to developing several goals that will enable you to accomplish each of your strategies. Goals should be SMART: specific, measurable, achievable, realistic and time-based.

In the example above, the HRAGD group might consider setting one goal to hold a monthly chapter meeting. Another goal that supports their strategies is to schedule a relevant seminar quarterly. Another goal might include holding informal dinners and cocktail hours to support voluntary member exchange.

Once you have enabled strategy accomplishment through setting SMART goals, you will want to develop action plans to accomplish each goal. Continuing with HRAGD as the example, to offer a quarterly seminar, you will need to follow an action plan:

- Establish a cross section of professionals as a committee and meet to plan the sessions.

- Determine budget.
- Perform HRAGD member needs assessment.
- Select topics based on member needs assessment.
- Locate exceptional speakers.
- Pick speaker and negotiate workshop length, pay, topic and objectives.
- Determine location and schedule the seminar.
- Plan advertising strategies, and so forth.

Make action plans as detailed as you need them to be and integrate the individual steps into your planning system. An effective planning system, whether it uses a personal computer, a paper and pen system, a handheld computer or a Palm, will keep your goals and action plans on track and on target.

Chapter 2

Food Production Basics

Food Hygiene Includes all Practices, Precautions and Procedures Involved in:

a. Protecting food from the risk of biological, chemical or physical contamination.
b. Preventing any organisms multiplying to an extent that would expose consumers to risk or result in premature decomposition of food.
c. Destroying any harmful bacteria in food by thorough cooking or processing.

Benefits from high standards of food hygiene include:

a. Reduced risk of food poisoning, foreign body contamination and spoilage.
b. Compliance with MOD and legal requirements.
c. Economic advantages, including increased shelf life and reduction of waste.
d. Consumer satisfaction and enhanced reputation.
e. Increased morale of personnel.

Hazard Analysis

The Food Safety (General Food Hygiene) Regulations 1995 (FS(GFH)R) refer to *Hazard Analysis* (HA). HA is based on five principles, detailed at Annex A to this chapter and are summarised below:

a. Analysis of the potential food hazards in a food operation.
b. Identification of the points in those operations where food hazards may occur.

c. Deciding which of the points identified is critical to food safety ("critical points").

d. Identification and implementation of effective control and monitoring procedures at those critical points.

e. Review of the analysis of food hazards, the critical control points and the control and monitoring procedures periodically and whenever the catering operation changes.

In order to meet the requirements of HA across the MOD, a practical guide for the MOD has been produced, based on the principle that there are four main methods of preparing food. The guide, *Hazard Analysis-The Four Line Method*, is at Annex B to this chapter, which is to be used as a template in all Service establishments.

Universally recognised Critical Control Points (CCPs) have been identified, however additional Control Points or Critical Points may be applicable in each case. This will be entirely dependant on the type of catering function and the supporting infrastructure. When preparing individual HA profiles, units are advised to consider the following definitions in order to determine the type of risk that applies at each stage of food preparation:

a. *Control point* – specific stages in those operations where food hazards may occur.

b. *Critical point* – control points which are considered critical points where the hazard must be controlled to ensure that it is eliminated or reduced to a safe level.

c. *Critical control point* – is a critical point in the process where a specific hazard must be controlled on the basis that no further process will adequately eradicate that hazard.

Written records are also important in establishing a Due Diligence defence under the Food Safety Act 1990. HA, as defined in the FS(GFH) R, includes the principles of: Critical Points and the Control of those Critical Points. In this document the term Hazard Analysis (HA) will be used, but this is be taken to be synonymous with HACCP.

Regulation 4(3) of FS(GFH)R states that: "a proprietor of a food business shall identify any step in the activities of the food business which is critical to ensuring food safety and ensure that adequate safety procedures are identified, implemented, maintained and reviewed".

This Regulation is designed to make caterers focus on the activities critical to food safety and to find ways of controlling them.

The systematic analysis of each individual food item would result in a heavy burden for most MOD catering operations and would be impractical to implement. An approach, that considers the catering operation step by step, from supply through to consumption, is achievable, effective and forms the basis of food HA in the MOD. The HA approach to ensuring food safety and hygiene is intended to give a clearer focus on the controls that are important to Service caterers to ensure that safe food is provided.

Where catering is provided by contract, the contractor is to use a Hazard Analysis system that meets legislative requirements or adopt the Four-Line Method. Specialist advice, through the Chain of Command may be obtained from DCG Service Support Contracts

Supply of Food and Water

The F(GFH)R state that no raw materials or ingredients shall be accepted by a food business if they are known to be contaminated with parasites, pathogenic micro organisms, toxic, decomposed or foreign substances and that, after normal sorting and/or preparatory processing procedures, they would still be unfit for human consumption.

Delivery Monitoring Routine checks must be made on deliveries of food for signs of damage, contamination and the presence of pests. The general condition of the food is to be checked, together with more specific checks such as date marks (Best Before and Use By) and temperature. Unfit food or food past its "Use By" date must not be accepted and is to be immediately returned. Temperature monitoring should be conducted at the time of delivery and should form part of the quantity & quality check.

For chilled and frozen foods, checks are to be made that the food is delivered at the correct temperature (less than 8oC for chilled and less than –12oC for frozen).

The Delivery Monitoring form is no longer required to be completed, the vehicle temperature printed record must now be attached to the Price Advice Note (PAN). For bulk stock "between pack" temperatures are to be taken using an air probe and recorded directly onto the PAN. A note stating "Quality & Date Coding Checked" is to be hand written onto the PAN. A commercially produced ink stamp could also be utilised for this task.

Potable Water Supply. It is essential that galleys and kitchens be supplied with potable water for the purposes of food preparation and cleaning.

Food Storage

The FS(GFH)R requires food products and ingredients to be stored in food premises in appropriate conditions, designed to prevent harmful deterioration and to protect food from contamination.

Dry Provisions Storage. These storerooms are to be kept clean and orderly to minimise the potential hazards from "foreign bodies" and to prevent the harbourage of pests. Where practicable these stores are to be proofed against pest ingress. Part used packs are to be adequately to prevent contamination. High ambient temperatures (above 13oC) and high humidity are to be avoided. Dry goods should be stored on suitable racking raised off the ground.

Chilled Storage (5oC). Cool rooms and refrigerators are to be kept clean and tidy. It is the responsibility of the Catering Manager to raise the appropriate request if refrigerated space is considered inadequate. Best Practice guidance indicates that the ideal operating temperature of a refrigerator is 1-4oC.

Frozen Storage(-18—21oC). Freezers are to be kept clean and tidy at all times. Best Practice guidance indicates that the ideal operating temperature of a freezer is between –18oC to –21oC.

Satisfactory storage is essential if a galley/kitchen/mess is to serve clean and safe food. This should include routine turnover of stock and checks of Best Before Dates. There are four main groups of foods that require differing storage conditions. The groups are as follows:

Where reference is made to a Priced Advice Note; this also means any other form of Invoice or Delivery Note

a. *Fresh Fruit and Vegetables.* These may be contaminated by soil bacteria and are to be stored away from other foods in a cool area, with adequate ventilation, preferably refrigerated at a temperature appropriate to the product.

b. *Dry Foods.* (including Canned Foods, Cereals and Flour). These are to be stored in dry, well-ventilated rooms ideally at 13oC. Food is to be put on racks, shelves or pallets off the floor.

c *Frozen Foods.* These are to be stored in a freezer as soon as they are delivered. Frozen foods are to be used within 1 month. 'Kitchen prepared' chilled foods that are subsequently frozen are also to be used within 1 month.

d. *Perishable Foods.* Perishable food temperatures in a refrigerator must not exceed the legal limit of 8oC. Therefore, perishable foods are to be stored in a refrigerator cabinet, set at a target temperature of 5oC. Such food includes chilled dishes, meat, poultry, eggs, cooked meats and vegetable dishes, prepared salads, soft cheeses, sandwiches, fresh pasta, low acid desserts, smoked or cured meats, fish and dairy produce. Best Practice advice, is that chilled foods are to be used within 48hrs. Food that is spoiled or past its "Use By Date" must be removed from food rooms and appropriate action taken prior to disposal. Broken packaging must be avoided.

Food Displays/Serveries. Food displays and serveries should be set at the correct temperature with adequate time allowed for the equipment to achieve target temperatures. It is good practice to have sneeze screens attached to hot and cold food displays to minimise the risk of contamination of food. The handles of utensils should not contact food; this can be achieved by using utensils with longer handles than the service containers.

Cross Contamination

Cross contamination is the process whereby pathogenic bacteria present in raw food, such as meat, poultry, and vegetables, come into contact with ready to eat food, either directly or indirectly, through contact with surfaces, equipment, splashes, drips, utensils, hands and cloths. To prevent cross contamination, raw and ready to eat foods are to be kept apart. Good management planning of workflow through the kitchen can assist in this. The flow of waste is also be considered, along with raw products and prepared food. Additionally, it is to be ensured that:

a. Where possible, there are separate designated preparation areas for raw and cooked foods. If, due to limitation of space, preparation surfaces are used for both high and low risk foods then these activities must be separated by time (ideally low risk foods prepared first). Preparation surfaces must be thoroughly cleaned and disinfected between the two operations.

b. If raw and cooked food is to be stored in the same refrigerator, raw foods are always to be stored (and adequately covered) below cooked or salad foodstuffs.

c. Food handlers always wash their hands between handling raw and cooked foods and change their protective clothing once heavily soiled.

d. Separate colour coded cutting boards (red for raw meat, blue for raw fish, green for fresh fruit, salad and vegetables, white for cooked foods) and colour-coded knives are used (details of NSNs for these items are contained in JSP 308). Always thoroughly wash and disinfect cutting boards and knives in between and after use.

e. Separate machines for raw and cooked food are used, or slicing and mincing machines are thoroughly cleaned and disinfected between raw and cooked food. If separate machines are not available, cooked food must be prepared before raw food.

f. All food is kept in covered, dated and labelled containers during storage and before service.

g. Storing and Cooking with Eggs. Some eggs can contain salmonella bacteria inside or on their shells so care must be taken with their use. They are to be stored in a cool dry place or refrigerated, with exterior packaging removed and kept away from other foods. All egg dishes are to be served immediately or cooled quickly and chilled. Commercially pasteurised egg products should be used in preparation of any recipe that is served uncooked or partially cooked. This includes mayonnaise; bearnaise and hollandaise sauces, some salad dressings, ice-cream, icings and tiramisu.

h. Washing of Fruits and Vegetables. All fruits and vegetables should be washed in potable water before use. In addition fruits and vegetables purchased locally overseas are to be immersed for 30 minutes in water containing either of the agents listed. The items should then be rinsed to remove taste of disinfecting agent:

   (i) Calcium hypocholrite granules-one level teaspoon per two gallons of water

   (ii) Milton: 5 teaspoons per gallon of water.

Defrosting Food

It is essential that frozen meat and poultry are thoroughly thawed on trivets, or if unavailable in a designated defrosting area, or ideally in a defrosting cabinet. Frozen vegetables must be cooked directly from frozen.

Cooking

It is essential that food is cooked thoroughly to destroy any bacteria on or within it. Food is to be cooked to a core temperature of 75°C, checked with a calibrated digital probe (disinfected with bactericidal wipes). It is important that food such as rolled joints, burgers, and chicken are cooked to 75°C at its thickest part.

Chilling Food

When chilling cooked food for later cold or hot consumption, it is essential that the food be cooled as quickly as possible. In accordance with 'Best Practice' hot food is to be taken through the recognised 'Danger Zone' of 60-10°C within 4hrs, and subsequently stored in a refrigerator at 8°C or cooler. (These requirements refer to the temperature of the food, not the air temperature of the equipment). If the target temperatures are not achieved food is to be discarded. The chilling of food to a recognised safe temperature may be achieved using the following methods:

a. Break food down into shallow trays and cover.

b. Cool initially to a low ambient temperature, then refrigerate or blast chill to below 'Danger Zone'.

c. Store in a refrigerator.

Reheating Food

It is advised that the reheating cooked foods is avoided, but if this is not possible it must only be reheated once and the following steps are to be taken:

a. Ensure that the food reaches a core temperature of at least 75°C. (In Scotland it is a legal requirement for the centre of the food being reheated to reach 82°C).

b. Use digital probe thermometers to check the temperature at the centre of the food. (Probes are to be disinfected using bactericidal wipes between each use).

c. The food is to be served and eaten as soon as possible.

d. After reheating, any leftover cooked food is to be thrown away.

Temperature Control and Monitoring

The Food Safety (Temperature Control) Regulations 1995 (FS(TC)R) impose two holding temperatures of below 8°C or above 63°C. The purpose of the regulations is to inhibit or prevent harmful micro organisms from multiplying by keeping food outside of the recognised 'Danger Zone' of 60°C to 10°C. The FS(TC)R allow limited periods outside temperature control during preparation, display, service, storage or transport, but it is an offence to keep food out of temperature control for so long that it could become unsafe. The requirements do not apply to foods, which by their nature or their packaging (e.g. canned foods) are inherently safe from the growth of such organisms. In such cases the manufacturer's guidelines are to be followed.

Refrigerator and Freezer Temperature Recording. The operating temperature of all appliances should be monitored and recorded thrice daily using either of the templates at Appendices 1 and 4 to Annex B of this Chapter. As a guide, the target operating temperature for a refrigerator is 5°C and for a freezer is-18°to-21°C, but this will depend on the efficiency, age, contents and capacity of the appliance.

Note: The template to be utilised is dependent on establishment size and the number of appliances.

Catering Standing Orders are to specify the corrective action (including Out of Hours/Weekend actions required) to be taken in the event of refrigerators/freezers not achieving recommended operating temperatures or failing completely.

Food Holding Temperatures

Two and Four Hour Rule

Hot food must be kept at 63°C or above. It must either be consumed within 2 hrs from the time the food temperature drops below 63°C, or be disposed of.

Chilled food may be displayed (away from under temperature control) for a maximum period of 4 hours. Only one such period is allowed, no matter how short. If, for example a refrigerated dish has been on display for 1 hour in a non-refrigerated environment, refrigerating it does not mean it can be displayed for a further 3 hours later. Where food is being cooled prior to refrigeration, cooling is to be achieved as quickly as possible as described in paragraph

When hot or cold food is displayed outside a temperature controlled environment, it is to be consumed or discarded.

Target Temperatures

Food Safety legislation and the Dept of Health require that certain temperatures be maintained throughout the food storage and production process.

- Cook
- Temp
- Service
- Temperature

- (At or above)
- Reheat Reheat
- Scotland
- Chilled
- Storage
- Frozen
- Storage
- Legally Required
- Temperature
- 0°C for 2 mins 63°C 75°C 82°C 8°C-18C.

Temperature Measurement and Temperature Probe Calibration

Measurement. A digital thermometer with attached probe is to be used for measuring food temperatures, including final cooking temperatures. The probe is to be disinfected with food safe wipes between measurements. There are also 'Between Pack Probes' are available that are to be used to measure temperatures of frozen and chilled foodstuffs, primarily at the delivery stage. Details of these items can be found in JSP 308.

Numbers of parts as available as follows;

Description	*NSN*	*Remarks*
Wipe Clean Thermometer	7320 99 0173073	
Robust Penetration Probe	7320 99 2154600	
Thermometer Between Pack Probe	6685 99 9906711	6 x 200 pack
Bacterial Probe wipes	7320 99 995890	
Pocket Stick Thermometer	6685 99 0007159	

Calibration: The digital thermometer is to be re-calibrated annually or in accordance with the manufacturer's instructions and warranty. A monthly calibration check is to be undertaken with an appropriate device in accordance with the manufacturer's instructions, and recorded on the template at Annex C to this chapter. The Test Plug is also to be calibrated in accordance with manufacturer's instructions and warranty.

Records

Records. Records are an important aspect of any due diligence defence and are essential in proving that the temperature requirements were being met in the event of any prosecution or food poisoning outbreak. Records (including Calibration records) are to be retained for 6 months. The mandatory records that are to be kept with unit files are listed below:

a. Delivery monitoring records.
b. All refrigerator and freezer temperature monitoring documents.
c. Food temperature monitoring records.
d. Calibration records for temperature probes.

Personal Hygiene

Good personal hygiene is a legislative requirement, ensuring safe food. Induction training must ensure that new staff joining are aware of the required standard Food handlers are a potential source of bacterial and physical contamination of food, and so personal hygiene is a key element in ensuring that food is prepared safely.

Staff training is to include all the basic elements of personal hygiene (including hand washing training) covered in this section and personnel are to understand the relevance of the precautions.

The FS(GFH)R lays down criteria for the personal hygiene of food handlers and the actions to be taken if a food handler is infected. These regulations require every person working in a food handling area to maintain a high degree of personal cleanliness.

Any person suffering from a disease that is likely to be transmitted through food must inform the person in charge of the galley/kitchen. A separate regulation states that any such infected person may not be permitted to work in any food handling area, in any capacity, where they might contaminate food.

Specific Requirements

a. *Hand washing.* The hands of food handlers are the principal agents in the transference of bacteria to food. Hand washing facilities (a separate basin that is only to be used for hand washing) in galleys/kitchens and heads/toilets are to include hot and cold running water, soap, nail brush and a suitable means of drying the hands. Regular checks by senior staff are to be made to ensure these facilities are available and are being used effectively. As a minimum, hand washing, is to take place:

(1) On entering a food room.
(2) Between handling raw and cooked food.

(3) After handling waste food or refuse.

(4) After smoking or eating.

(5) After visiting the WC.

b. *Cuts.* Open cuts harbour bacteria and must be covered with clean blue waterproof dressing to aid detection. Stocks of these are to be available in food preparation areas and readily accessible to food handlers.

c. *Jewellery.* The wearing of wristwatches, ear rings ('Studs' are acceptable) and other exposed body piercing, bracelets and rings, is not acceptable as they harbour bacteria and there is a risk of physical contamination of food. The wearing of a plain wedding ring is acceptable.

d. *Smoking.* Smoking is not permitted in food areas, as it transfers bacteria from the mouth to hands. Cigarette ends and ash also pose physical contamination risks.

e. *Protective Clothing and Changing Facilities.* Food handlers must wear suitable clean protective clothing (including appropriate footwear and hats) to prevent contamination of food from normal clothing.

Such clothing is to be changed at the end of a shift, or sooner if the situation requires, in order to maintain hygienic standards. It is not to be worn outside food areas and associated premises.

If necessary, an outer garment is to be worn over protective clothing whilst away from food production environment. Adequate changing facilities must be provided, with locker space for clean and soiled protective clothing. Changing facilities must be kept clean and tidy at all times.

f. *Visitors.* All visitors to a food preparation area are to be viewed as potential sources of contamination. They must therefore be provided with protective clothing and briefed upon good food hygiene practices before entering the food area. Before entering a food preparation area, visitors must confirm that they are not suffering from diarrhoea and/or vomiting, or heavy cold. Individuals who are suffering should not be allowed to proceed.

g. *Eating/drinking.* Food handlers are not to eat or drink in food rooms. It is acceptable for cooks to taste dishes during preparation in a manner that does not contaminate the food i.e. a clean spoon each time.

h. *Toilet Facilities.* Where possible toilets for catering personnel are to be separate from those for non-catering personnel and visiors. Toilets are not to be located directly within a food preparation area. There is to be an intervening ventilated space between toilets and food rooms. Food is not to be stored in that space. Toilets must be ventilated such that associated odours are prevented from permeating into food rooms. They must also be kept clean and tidy, in good repair with adequate supplies of toilet paper, hand washing facilities, nail brushes, soap and hand towels.

Infected Food Handlers

Dispersed Feeding

Dispersed feeding is the production and transportation of food for consumption away from an established unit/galley/kitchen or mess for example to personnel on security duties who are unable to be fed in-mess, non-public section functions, such as barbecues or sports meals. This method of feeding is recognised as high risk, because the majority of foods used require controlled temperatures during transportation in order to comply with Food Safety Legislation. Dispersed feeding should only take place when there is no suitable alternative.

Documentation. Before food leaves the kitchen, the Shift NCO (or equivalent if DEL or contract catering staff) is to ensure the completion (in duplicate) of FCAT 1013.

Dispersed Feeding Record(demanded from DSDC Llangennech)

A copy is to be retained with the food safety records and a copy handed to the person (recipient) taking the food out of the kitchen. The recipient is to be briefed on the following points:

a. Cilled Products. Chilled foods are to be consumed within 4 hours of being removed once cold products leave from a temperature controlled environment i.e. refrigerator or chilled servery and when and the product temperature rises above 8oC, products are to be consumed within 4 hours.

b. Hot Products. Hot foods are to be consumed within 2 hours of being placed into a container. Food is to be served at 63°C. If a hot container meal arrives at its final destination and the product temperature is below 63°C, it is to be consumed as soon as possible (but always within 2 hours from whe the product was placed into the container). This time will have been recorded in column (b) of the FCAT 1013.

c. When temperature monitoring is unavailable at a dispersed feeding location, the food is to be consumed within 2 hrs (hot product) or 4 hours (cold product) of being placed in the insulated container or leaving a refrigerated environment. Once again the time of service is to be recorded in column (b) of FCAT 1013.

Transport Of Food

Movement of temperature-controlled foods may include:

a. The collection of food from suppliers and transportation to the main ration stores and messes.

b. The transportation of prepared food from messes to satellite or dispersed feeding facilities.

Transport can include items such as trolleys, bags, boxes, trays and crates. These articles must be kept clean. Where transport is also used for materials other than food, it must be thoroughly cleaned between loads to avoid the risk of cross contamination.

Non-catering personnel collecting foodstuffs from the mess are to be briefed on the importance of maintaining the integrity of the food chain, and how this is achieved.

During transport, all food is to be covered and is to be transported in appropriate insulated containers. High-risk food must be kept separate from anything that could cause contamination. When transporting prepared meals, dispatch and receipt temperature checks are to be recorded in accordance with paragraph.

Cleaning

The FS(GFH)R requires adequate facilities are provided for the cleaning and disinfecting of work utensils and equipment. These facilities must be constructed of materials resistant to corrosion, must be easy to clean and have an adequate supply of hot and cold water.

Equipment items that come into contact with high-risk foods will need to be cleaned and disinfected as necessary. Facilities must be provided to clean and disinfect all tools and equipment, crockery, cutlery, glasses and serving dishes. There are to be sufficient cleaning materials and equipment to match the size of the catering facility. Staff involved in cleaning must wear suitable and adequate protective clothing, which must be changed, at the end of each shift.

Colour coded cleaning equipment, e.g. mops must be cleaned, dried and stored in a designated area once finished with. Drying of equipment must not cause recontamination, e.g. from soiled cloths. A full list of NSNs of colour-coded equipment is shown in JSP 308.

Equipment must be of durable construction and resistant to corrosion, especially items that will come into contact with powerful cleaning chemicals.

Sinks. It is a legal requirement that, where appropriate, adequate provision must be made for any necessary washing of food. Every sink or other facility for the washing of food must have an adequate supply of hot and/or cold potable water as required, and be kept clean. Separate facilities must be provided for hand washing, food preparation and equipment washing if the volume of preparation in the kitchen demands it. In small catering operations, one sink may be used for both equipment and food washing, provided that both activities can be done effectively and without prejudice to food safety, i.e. the sink is to be thoroughly cleaned between each process. It is good practice to have signs above sinks indicating what they can be used for. Hot water supply is not essential if a sink is to be used exclusively for food preparation.

Disinfection. Disinfection is a process that reduces microorganisms to a level that will not lead to harmful contamination or spoilage of food. Sterilisation is a process that destroys all living organisms. The word 'sterilise' is sometimes used in food hygiene literature and in some official publications where the word 'disinfect' would be correct. Cleaning products are often described as 'detergent/sterilise' when 'detergent/disinfectant' would be a more accurate description. The following definitions clarify the relationship between disinfection and cleaning:

a. A disinfectant is a chemical agent used for disinfection after cleaning.
b. Sanitisation is a term used mainly in the food and catering industry. It is a process of both cleaning and disinfecting utensils and equipment.
c. A sanitiser is a chemical agent used for sanitisation.
d. Sterilisation is a process intended to destroy or reove all living organisms. It is not to be confused with disinfection.
e. Cleaning, Heat, and Chemical Disinfectants. Disinfection is essential in food hygiene. It is achieved by:
   (1) Cleaning. This is the most useful and practical method of removing food residues, dirt, grease and other undesirable

debris. It can be used alone or in combination with the other two methods. A high proportion of microbial contamination is removed from equipment, food preparation surfaces and hands by thorough cleaning.

(2) Disinfection by Heat. This is considered the most reiable of the three methods of disinfection. Immersion in water at 65°C for 10 minutes, or at higher temperatures for shorter times can be relied upon to destroy most microorganisms harmful to health, with the exception of bacterial spores. Boiling water is an effective and cheap disinfectant although bacterial spores may survive boiling. The temperature of the washing or rinse cycle in a dishwashing machine can be adjusted to include heat disinfection in the cycle, in accordance with the manuacturer's instruction.

(3) Some chemical disinfectants are active against a wide range of bacteria; others have a narrow range of antibacterial activity. In all circumstances chemical disinfectants must be used in accordance with the manufacturer's instruction. It is military policy to use general-purpose detergent for general cleaning with surface sanitisers. Chemical disinfectants (stored in a designated

Cleaning Procedures

Cleaning of food premises is carried out to remove dirt and grease from all surfaces and equipment, including pathogenic bacteria, that could compromise food safety, as well as food debris.

It is imperative that high standards of cleaning are maintained for the following reasons:

a. To comply with the legal requirements of the FS(GFH)R.
b. To reduce the risk of food poisoning by removing food residues, which could contain harmful bacteria.
c. To deny pests harbourage and food.
d. To reduce the risk of foreign objects physically contaminating food.
e. To promote hygiene awareness amongst catering personnel.
f. To provide a pleasant working environment and a favourable image to personnel.

Cleaning Schedule: A planned cleaning schedule programme is important in ensuring that high standards of cleanliness in all food areas are achieved and maintained. This can be achieved by adhering to a written cleaning schedule. Food preparaion surfaces must be regularly cleaned whilst in use. This is particularly important where space is limited and the cross contamination risk is increased.

a. Instructions for cleaning specific areas or pieces of equipment are to include the following information relative to the task:
   (1) Job description.
   (2) Cleaning materials and chemicals to be used (in accordance with manufacturer's instruction).
   (3) Safety precautions.
   (4) Job method.
b. The cleaning schedule is to state how often a specific area or piece of equipment is to be cleaned and who is responsible for checking that all cleaning tasks have been completed to a satisfactory standard.
c. General information concerning the cleaning of food equipment, surfaces and the material structures, including a list of cleaning equipment and agents used in food rooms/kitchen/galley/mess, are at Annex D to this chapter.
d. Catering managers are to implement a cleaning schedule that relates to all food areas within their department. A schedule detailing the frequency of routine cleaning tasks should be produced for each catering facility. It is to be contained within Catering Standing Orders and communicated to all personnel (e.g. by displaying the schedule on a notice board). A matrix showing the frequency, details of the task, type of cleaning required and a signature block is considered the most useful layout and should be readily available for operatives to follow and sign off. A signature block should also be included for a supervisory check.

Deep Cleaning: All galleys/kitchens and associated areas are to be deep cleaned in accordance with single Service instructions. Catering managers are to ensure that deep cleaning contracts are adequate to meet the tasks required.

Food Preparation Areas

Surfaces. The FS(GFH)R require that those surfaces which come into contact with food must be maintained in a sound condition, be easy to clean and, where necessary, disinfect. This requires the use o smooth, washable and nontoxic materials. To comply with this legal requirement, it is important that surfaces that come into contact with high-risk foods must be capable of being disinfected regularly. Examples of such surfaces, assuming that they are properly fixed, applied or installed and maintained, include: stainless steel, ceramics, and food grade plastics. Wooden boards are inappropriate for the preparation of high-risk foods. Food preparation surfaces must either be continuous in their construction or have properly sealed joints.

Design and Layout. The design and layout of food areas are to:

a. Permit adequate cleaning and/or disinfection.

b. Protect against dirt, contact with toxic materials, and formation of condensation and mould.

c. Encourage good food hygiene practices and prevent cross contamination by foodstuffs, equipment, materials, water, air supply or personnel and external sources of contamination such as pests.

d. Where practicable, suitable temperature conditions should be provided for the hygienic preparation of food.

Structural Requirements. In areas where food is prepared, consideration is to be given to floor and wall surfaces, ceilings and overhead fixtures, windows, doors and other openings. There are to be adequate washbasins, sanitary conveniences, ventilation, and drainage. Any maintenance requirements and requests to catering premises are to be recorded in either the Maintenance of Catering Premises – Record of Work Services Template at Annex E to this chapter or the unit works service record. Priority awarded to work service request is entirely dependent on the type of establishment and the impact of the equipment failure. The following priorities are to be applied:

MOD Standards. Defence Estates (D) Design and Maintenance Guide 18 deals with the design of kitchens and serveries for all ranks, and dining rooms for Junior Ranks. Details of dining rooms for Officers and SNCOs are given in JSP 315 Services Accommodation Code, Scales 29 and 34. The guide is also to be read in conjunction with DE Specification 42 on Catering Equipment and JSP 315, Scales 01, 39, 40, 45, 47 and 52. For HM Ships and Submarines, the relevant Naval Engineering Standards (NES) publication provides guidance in the design, layout and equipping of Galleys, Storerooms (refrigerated and dry) and associated areas.

Pest Control

Pests are known to carry a number of pathogenic organims that can be transmitted to humans through contaminated food. In addition, pests will damage food stocks causing financial loss. It is therefore important that food premises are kept pest free. The Prevention of Damage by Pests Act 1949 and the FS(GFH)R impose legal duties on owners and occupiers of buildings to keep their premises free from infestation.

A wide variety of insect and rodent pests will enter food premises for a number of reasons:

a. Food. Even in small quantities, food will enable pests to survive and multiply. Regular and thorough cleaning of spillages is therefore imperative.
b. Warmth. Pests of all types are attracted to buildings, which offer even limited warmth away from outdoor conditions. A few degrees increase in temperature will provide conditions in which breeding is enhanced and proliferation encouraged.
c. Shelter. Almost every building provides a variety of harbourages for pests.

Contrary to common belief, it is the newer buildings with suspended ceilings, panelled walls, service ducts and enclosed electrical trunking, which are more likely to create a problem, than older buildings without such features. Access must be provided to these spaces for the effective control of pests.

Denial of food, warmth or shelter will prevent the survival of pests. This form of control can be termed 'environmental control' and is the first line of defence against possible infestation. Environmental control may be considered as denial of access (proofing), food and harbourage. Flying Insect Control. Emphasis is to be placed on the environmental and physical control detailed previously to reduce the risk of food contamination. Areas around food premises are to be kept clean and tidy to reduce the number of possible breeding sites.

External refuse containers are to be clean and in good repair, and have tight fitting lids. If skips are used they are to be completely enclosed. Waste food containers must be washed out before being stored outside. Windows and other openings which provide ventilation are to be fitted with close fitting and cleanable fly screens where required. Doors are to be kept closed or fitted with screens or clear plastic heavy-duty strips. Electronic fly killers, where required, are to be installed but must not be sited above food preparation areas. The bulbs within electric fly killers must be changed in accordance with the manufacturer's instructions.

Recording Sightings

All personnel employed in a kitchen/galley have a duty to report to the Head Chef any evidence of pest infestation. Kitchen Managers are to record such sightings in the Pest Management Register at Annex F to this chapter. Under operational conditions or on exercise, trained Service personnel carry out pest and vector or vermin control. In peacetime (except in some overseas Commands and HM Naval Bases where MOD Civilian pest control operators are employed) civilian contractors normally carry out pest control. There are two ways in which a pest control contractor may be used:

a. To deal with and eradicate a single infestation.

b. To act as a long-term contractor who will visit the premises regularly and carry out pest control treatments as necessary. This proactive approach is more suitable and is recommended.

Waste Disposal

The FS(GFH)R require that food waste and other refuse must not be allowed to accumulate in food rooms, except so far as is unavoidable for the proper functioning of the catering department. It is recommended that systems of work are in place to ensure that refuse containers in food rooms are not over filled and are emptied regularly. All waste is to be removed at the end of the working day.

Food waste and other refuse must be deposited in sealed containers which must be of an appropriate construction, kept in sound condition, and, where necessary, be easy to clean and disinfect.

Lids are not required on refuse containers used for temporary storage of waste in food preparation areas. They are frequently touched by the hands of food handlers and may be a serious source of contamination. Any refuse containers used for storage of waste awaiting collection and removal are to have a lid and to be constructed of durable material, which makes them easy to clean and disinfect. Refuse containers areto be included in the cleaning schedule.

The FS(GFH)R state that adequate provision must be made for the removal and storage of food waste and other refuse. Refuse stores must be designed and managed in such a way as to enable them to be kept clean, prevent access by pests, and protect against contamination of food, drinking water, equipment or premises.

Areas for indoor storage of refuse must be remote from food rooms and not sited near the main delivery entrance. In establishments it is good practice to have a separate area designated for the storage of outdoor waste with well-lit hard standing. A hose is to be provided for cleaning purposes.

Environmental Protection (EP) and Marine Pollution Regulations (MARPOL)

Disposal of food waste must meet the requirements of EP (JSP 418), DEFRA instructions and MARPOL legislation. Revised legislation relating to the disposal of raw meat and fish produced as waste product came into effect 01 Jul 03. Such products can no longer be sent to landfill for disposal. A transitional period until Dec05 has been granted before which DCG will publish policy regarding disposal of such products.

The Five Principles of Hazard Analysis

Analysis of the Potential Food Hazards in a Food Business Operation

Food Hazard: A hazard is anything that could cause harm to the consumer. There are three main hazards that may arise with food served in catering premises. These are contamination of food by:

- Bacteria or other microorganisms that cause food poisoning.
- Chemicals, for example by cleaning materials or pest baits.
- Foreign materials such as glass, metal, plastic and so on.

Of these, the most imporant hazard is likely to be harmful bacteria or other germs that may contaminate and multiply in food. Identification of the Points in those Operations where Food Hazards may occur. Food passes through many steps; delivery, storage, preparation, cooking, cooling and so on. Hazards can occur at many or all of the steps. For each type of food, the hazards that may occur at each step are to be identified. Mostly these will be steps in the operation where:

- Food can become Contaminated with microorganisms, chemicals or foreign material.
- Bacteria can Multiply if the food is held too long at the wrong temperature.
- Micro organisms Survive a process that should kill them, e.g. when cooking or disinfection of equipment is inadequate.

Deciding which of the critical points identified are critical to ensuring food safety.

Critical Points are steps at which the hazards must be controlled to ensure that a hazard is eliminated or reduced to a safe level.

- Any step where food may be become Contaminated is to be controlled. Controls include clean and disinfected equipment, the personal hygiene of staff, and separation of raw and cooked food. All food is to be protected from contamination by foreign bodies, pests or chemicals.
- Steps where bacteria may be able to Multiply in food must be controlled. The time and temperature at which food is held, stored or displayed are likely to be critical.
- Any cooking or reheating step is to be able to Kill harmful microorganisms. It will be critical that heating is thorough. Cooking is normally the most important control step in most food preparation. Chemical disinfection of equipment is another control point designed to Kill microorganisms.

Identification and implementation of effective control and monitoring procedures at those critical points

Controls must be set for the Critical Points, then Checks Introduced

Controls will either reduce the hazard to an acceptable level or get rid of it completely. The controls are to be precise as possible.

For example, it is better to state that certain products will be stored under refrigeration at a set temperature, rather than to simply say that it must be kept in the chiller. A control target is to be set for every critical control point that has been identified. When controls have been set, it is then possible to Monitor the critical points whenever that preparation step is used. The targets can be checked.

The frequency of checks is to be set for each control. Checking temperatures does not always involve probing food with a thermometer. Delivery vans or storage chillers may be fitted with temperature monitors and these can be checked. Other critical controls are more difficult to measure, e.g. cleaning & disinfection of equipment or the personal hygiene of staff. They will be often vital to the safety of food, and there are to be regular checks that standards are kept up.

Review of the analysis of food hazards, the critical control points and the control and monitoring procedures periodically, and whenever the food business's operations change. It is not satisfactory to go through this process once and forget about it. Occasionally the system may need to be reviewed and amended, for example:

- If controls or methods of checking are found to be ineffective or impracticable.
- The menu changes. New dishes may have new hazards and controls.
- The method of preparation changes.
- New equipment is introduced.

Annex B-Hazard Analysis

The Four-line Method

This annex contains:

a. Flow Diagram-An Overview of the Four-Lines.

b. Flow Diagram LINE A – Food/dishes served cold.

c. Flow Diagram – LINE B – Food/dishes served hot.

d. Flow Diagram – LINE C – Food/dishes cooked, chilled and served cold.

e. Flow Diagram – LINE D – Food/dishes cooked. Chilled and reheated.

f. Hazard Analysis Flow Chart – LINE A.

g. Hazard Analysis Flow Chart – LINE B.

h. Hazard Analysis Flow Chart – LINE C.

i. Hazard Analysis Flow Chart – LINE D.

j. Galley/Kitchen/Mess Management Records:

   (1) Daily Food Safety management Record (Template)

   (2) Advance Food Preparation Record (Template)

   (3) Food Time/Temperature Record (Template)

   (4) Galley/Kitchen/Mess Fridge/Freezer/Blast Chiller Daily Monitoring Record (Template)

   (5) Galley/Kitchen/Mess Fridge/Freezer/Blast Chiller Monitoring Record (Template).

Use of the Four-Line Method

All caterers are to use this guide as a reference document. Each catering establishment will have unique hazards, which are to be annotated onto the template as a Control Point, a Critical Point or a Critical Control Point.

The Four-Line Method is to be used as follows:

a. Monitoring and Recording of the HAL's. Record the daily menu on single Service record forms. Determine high-risk menu items and, by referring to the Four-Line method guidance, annotate on the record the relevant Hazard Analysis Line (HAL) to the individual food dish.

b. Whilst the Four-Line method provides a system of HAL within kitchen and provision areas, other factors will need to be considered in ancillary kitchen areas e.g. delivery bays and pot/crock wash.

Food Hygiene in Catering Establishments

Legislation and Model Regulations

Food and Agriculture Organization of the United Nations

The World Health Organization, through its Food Safety Programme, which is aimed at ensuring the safety of foods for the consumer, initiated a survey of legislation related to food hygiene in catering establishments. The results of the survey, which are summarized in this publication, and the model food hygiene regulations for catering establishments, also included here, were reviewed by a joint FAO/WHO expert consultation held in Geneva from 28 April to 2 May 1975.

This meeting was funded by the United Nations Environment Programme (UNEP) as part of its activity to "support, and accelerate or expand the work of the FAO/WHO Codex Alimentarius Commission on international standards for pollutants in food and strengthening of FAO/WHO capabilities to assist developing countries in food control". In this activity the Food and Agriclture Organization of the United Nations and the World Health Organization are cooperating agencies.

Many of the legislative texts on which this survey was based have been published, in full or in summary, in:

(1) International digest of health legislation. Geneva, World Health Organization (published quarterly); and

(2) Food and agricultural legislation. Rome, Food and Agriculture Organization of the United Nations (published quarterly).

The survey covers a reasonably representative cross-section of the legislation of Member.

States in all Parts of the World

Food control has two principal objectives: protecting consumers against health risks, fraud, and adulteration of foods and assuring fair practices in the food trade. The survey described here was concerned with the first of these objectives, namely, the safety and cleanliness of food, i.e., food hygiene.

The need for a high standard of food hygiene and for adequate control measures is particularly important in catering^ establishments of all kinds. The fourth report of the WHO Expert Committee on Environmental Sanitation-' points out that:

> *"One careless food handler, or one human carrier of disease, preparing food at home will jeopardize the health of only a small number of persons, mainly members of the family. When one such person works in the kitchen of a restaurant, hospital, factory, canteen, school, or other place where meals are supplied to many people, the number of potential victims is correspondingly greater."*

The present report is therefore concerned essentially with the promotion of food hygiene in catering establishments.

Nevertheless, since the basic principles of food hygiene are common to all types of food establishments and food handling activities, the model regulations annexed to this report can be used as the basis for the legal control of all types of food "Food hygiene" has been defined as, "The measures whereby the wholesomeness, soundness and safety for human consumption are secured or increased, covering all facets of food production, harvesting, processing, distribution, preparation and service and of possible causes of toxicity (physical, chemical or microbiological)".

Geneva, World Health Organization

The model code sets out in detail, but in simple terms, the various points which should be legally enforceable in any country and should be embodied in regulations or secondary legislation. It is intended, in particular, for the guidance of developing countries wishing to build up an effective organization for the control of standards of hygiene in catering establishments. In drawing up this code account has been taken of a statement by the Codex Alimentarius Commission on general principles of food hygiene.

Measures for Promoting Food Hygiene

Effective legislation is essential in any adequate system for promoting food hygiene, but this is only part of such a system. However comprehensive the legislation, it will be ineffective unless it can be enforced. This means that there must be an agency, or agencies, charged with that responsibility. These responsible bodies must in turn employ an adequate number of capable and well trained staff for enforcement duties. Indeed, one of the first steps which developing countries wanting to build up a service to improve standards of food hygiene should take is the development of a trained field staff.

Field staff cannot, however, be engaged solely in law enforcement activities. Their principal role must be to exercise constant surveillance over catering establishments so that they are fully aware of the standards of hygiene being observed and are able to take prompt action to correct any faults that they discover. This work will involve frequent inspections of premises and the collection of samples for bacteriological and chemical analysis, and supporting services such as laboratory facilities will be required. The field staff will also have advisory and educational functions, obtaining improvements in the handling of food through persuasion and consent and the education of managements.

To establish and maintain an adequate field staff with the support services required and to meet related expenditure on, for example, transport, equipment and materials, will require adequate financial resources. Part of the finances required may come from licensing or registration fees or other charges on the catering industry, but the major part will usually have to be obtained from central or local governments, or both.

Governments have to decide how the funds are to be provided but it may be necessary to present strong arguments to justify the request, particularly where a country has previously not had a comprehensive food control service. Since the primary object of a food hygiene programme is to protect the health of the public, it is not always easy to present a case on a cost-benefit basis. It must be recognized that promotion of food hygiene is only one of a number of measures for protecting and improving human health that fall within the scope of environmental health. According to a WHO Expert Committee on National Environmental Health Programmes: Their Planning, Organization, and Administration, environmental health "refers to the ecological balance that must exist between man and his environment in order to ensure his well-being".

Within the field of environmental health, environmental sanitation was defined by a WHO Expert Committee in 1950 as "the control of those factors in man's environment which exercise or may exercise a deleterious effect on his physical, mental, or social well-being".

The report of Joint FAO/WHO Food Standards Programme, Codex Alimentarius Commission. Recommended international code of practice: general principles of food hygiene. Rome, 1969.

WHO Technical Report Series, No. 439, 1970.

Committee on Environmental Sanitation). the WHO Expert Committee on National Environmental Health Programmes lists 17 item considered to be included in, or related to, environmental health. Among those items are water supplies, wastes disposal, vector control, and food hygiene.

Food hygiene, therefore, is an integral part of a health service and it is important to appreciate that it is difficult to develop in isolation a satisfactory programme for raising standards of food hygiene. High priority should be given to achieving proper standards in this particular field, but progress will be made more easily if efforts are also made to improve general environmental health.

In other words, a food hygiene programme should be developed within the context of progressive environmental health and personal health programmes. Good food hygiene is impossible without adequate supplies of safe water, proper means for wastes disposal, effective pest control, and reasonable living conditions. It is also difficult to reduce the incidence of food-borne infections where disease is widespread among the population and adequate steps are not taken to reduce sources of infection.

Current Legislation

Registration and Licensing of Premises

To exercise control over premises used for food handling the responsible agency must know the location of the premises, and its task of ensuring that these premises are properly constructed and adequately equipped, and that satisfactory hygienic standards are maintained, is facilitated if a business is not permitted to open until the premises are officially approved. This power of control is strengthened if the agency is able to enforce the closure of unhygienic premises.

Many countries recognize the need to control the opening of catering premises so as to ensure that before business commences they comply with the standards in force. Provision is usually made also for the closure of premises in the event of serious contraventions of the law.

The opening of premises is controlled in various ways. Sometimes a permit or licence from the local health department is required, and this may be revoked or suspended if the premises are allowed to fall into an unsatisfactory condition. One large city requires an applicant for a permit to complete successfully a course of instruction in food protection conducted by the health department. Sometimes there is a system requiring premises to be approved before they are opened.

In some countries control is achieved by a requirement that all catering establishments must be registered. For practical purposes no real distinction can be drawn between licensing and registration although licensing is normally for a limited period and implies periodical renewal of licences. However, the health authority will need for its own purposes a recordof the premises for which it is responsible, and a registration scheme therefore seems most appropriate.

WHO Technical Report Series, No. 439, 1970

Registration of food establishments is the process by which food control authorities, local or national, maintain registers in which is entered information that applicants must supply before they may operate. This information should include the name and address of the establishment and other relevant data. The approval of premises may be required before registration; however, this practice is not universal. Registration is usually for an indefinite period of time, subject to cancellation for causes indicated in the law. Licensing of food establishments is the process by which appropriate food control uthorities, local or national, issue licences to operators of food establishments that have complied with relevant legal requirements. The licence authorizes the establishment to operate for a specified period of time, after which renewal is required. Licences may be suspended or revoked as provided for in the law.

Where licensing or registration schemes are not operated the task of the health authorities is more difficult, but they could still have power to enforce the closure of premises that present a danger to health. The requirement that premises may not be opened for business until they have been registered or licensed is of value to both the health authorities and the operator of the business. If catering premises may be opened without any prior consultation with the health authority, or without their approval, the operator may be faced subsequently with the need to make substantial alterations at heavy expense in order to comply with the food hygiene regulations. Such expense might be avoided if the operator-is required to consult the health authorities at the outset and accepts guidance from health officers in the design and equipping of the premises. Indeed, the health authorities could warn operators against attempting to use an unsuitable building that could not be. made to comply with the required standards.

Authority to enforce the closure of insanitary premises is most .important. Clearly, there must be a means of speedily closing catering establishments that constitute a danger to public health, but the machinery for doing this requires much consideration.

Construction

Regulations for the construction and equipping of catering premises in force in different countries vary considerably in detail although their basic intention is to provide conditions under which food can be prepared hygienically and without endangering the health of the consumer. There can be little dispute about the basic constructional elements required, or about the essential pieces of equipment necessary for good food hygiene. It is obvious, however, that there is a wide difference of opinion on the extent to which it is necessary to impose on operators of catering businesses detailed legislative requirements concerning the buildings in which they carry on their business and the equipment to be installed.

Regulations may, in addition to specifying some matters in detail, state certain broad principles. For example, one set of national regulations for food premises contains this provision:

"Every food premises shall be so constructed, located and maintained that,

(a) the premises are free from every condition that may,
  (i) be dangerous to health,
  (ii) injuriously affect the wholesomeness of food prepared, processed, packaged, served or stored therein; ..."

Some other regulations have a similar general requirement worded as follows: "No food business shall be carried on at any insanitary premises or place or at any premises or place the condition, situation or construction of which is such that food is exposed to the risk of contamination."

A Further Example Reads

"Any person who sells, prepares, packages or stores for sale any food under insanitary conditions shall be guilty of an offence." General provisions of this kind are desirable because in legislation it is difficult to specify in detail all conditions which should constitute an offence. The extent to which legislation contains detailed requirements regarding the construction of catering premises varies considerably.

The points usually covered are that in rooms where food is stored or prepared the floors must be of smooth, nonabsorbent materials and so constructed as to be easily cleaned; all walls and ceilings in such rooms must be capable of being readily cleaned; all parts of these rooms must be kept in good order, repair and condition so as to enable them to be effectively cleaned. Premises are often required to be of such construction and in such a state of repair as to resist the entry of rats, mice and insects. In some causes, openings to the outer air are required to be effectively protected against the entry of flies and other flying insects.

Lighting and Ventilation

It is usual to require all food preparation and storage rooms to be adequately lighted and ventilated. This is normally stated in general terms but in some areas the degree of illumination is specified in detail. For example, one set of rules prescribes that at least 215 lux (20 foot candles) of light must be provided on all working surfaces and equipment in food preparation, utensil washing, and hand washing areas and in toilet rooms. At least 54 lux (5 foot candles) at a distance of 75 centimetres (30 inches) are required in dining rooms and all other areas during cleaning operations.

Equipment and Utensils

It is usual for food hygiene legislation to contain provisions concerning equipment and utensils. Legislation may require equipment and utensils to be of such materials, workman-ship and design as to be smooth, easily cleaned and resistant to damage. Provisions are sometimes laid down that all surfaces that come into contact with food must be readily accessible for cleansing and inspection. It is quite common to impose an obligation to ensure that all articles, equipment and surfaces must be nontoxic and, as far as possible, nonabsorbent. In one country legislation requires equipment and utensils to be:

(a) of sound and tight construction;

(b) kept in good repair;

(c) of such form and material that it can be cleaned and disinfected;

(d) free from cracks, crevices and open seams; and

(e) corrosion-resistant and nontoxic.

A most important part of the operation of catering establishments is the cleansing of equipment and utensils. The duty of ensuring that these are effectively cleansed after use and maintained in a clean and sanitary condition may be imposed in very general terms, leaving it to the operator of the business to adopt whatever methods he regards as the most effective. In such cases the legislation may simply require facilities for the cleansing of utensils and equipment to be provided and utensils and equipment to be thoroughly cleansed after use and kept clean. O the other hand, regulations in force in some countries specify in considerable detail how the cleansing of utensils and equipment should be carried out.

Sometimes, sinks with two or three compartments must be provided if cleansing is performed manually. Such detailed regulations usually divide cleansing into three operations: prerinsing, cleansing, and disinfecting or sanitizing. The temperature of the water used for cleansing may be laid down, 43.5°C (110°F) being usual. For machine washing, a higher water temperature may be specified.

Use of a Detergent may also be Obligatory

There may be a requirement that after utensils have been washed they should be sanitized by immersion either in hot water or in water containing a sterilizing agent. The temperature of the water used for sanitizing is usually required to be between 76.5°C and 82°C (170°-180°F) with an immersion time of 30 seconds at the lower temperature or 10 seconds at the higher. Where methods of chemical sterilization are specified the required immersion time varies from 45 to 60 seconds. The sterilizing agents allowed are solutions containing 50-100 parts per million of available chlorine, 12.5-25 parts per million of available iodine, or quaternary ammonium compounds at a concentration of not less than 200 parts per million. These solutions are to be used at a temperature of not less than 24°C (75°F).

Rules are also sometimes laid down for the operation of dishwashing machines. Where these are of the immersion type the. requirements are similar to those for manual dishwashing. For spray-type machines the temperature of the rinsing and sanitizing water may be specified. Detailed instructions may be given for cleansing procedures. These may require that, before being washed, articles must be flushed or scraped, and when necessary soaked, to remove large food particles and soil. Effective concentrations of a suitable detergent may be insisted upon for both manual and mechanical dishwashing.

As a check on the efficiency of. dishwashing, utensils may have to meet a prescribed bacterial standard. This is usually a plate count of not more than 100 bacterial colonies per utensil, using the swab technique. In the design of legislation for the cleansing of utensils and equipment there are clearly two different philosophies.

The first is that it is sufficient merely to require the provision of adequate facilities for washing, such as hot and cold water and sinks, and to make it obligatory to cleanse all vessels and other equipment thoroughly after use.

The operator of the food business is left to decide for himself the precise methods to be used to achieve a satisfactory standard of cleanliness. The second philosophy is that not only must there be an obligation to provide the necessary cleansing facilities, but the methods to be used must be prescribed in considerable detail including the temperature of the hot water and procedures for sterilization.

This philosophy also maintains that visual cleanliness is not enough; after being washed and sterilized, equipment must satisfy a bacteriological test. The need for strict cleanliness is unquestionable and the law should be adequate to enforce this. It is doubtful, however, whether there is really any need to embody in legislation the precise methods and materials to be used. If failure to cleanse equipment adequately is made an offence with an appropriate penalty for noncompliance, this should give the health officer the necessary authority to ensure that proper cleansing procedures are followed. His training should enable him to advise on how the required standards can be maintained.

Washing Facilities for Staff

All food premises need adequate washing facilities for the staff. They must be provided with a sufficient number of suitable wash-hand basins for the use of all persons engaged in the handling of food on or around the premises, and the basins should be conveniently placed in an accessible position. Wash-hand basins should be provided with an adequate supply of hot and cold water or of hot water at a controlled temperature. Soap or other suitable detergent, nailbrushes, and clean towels or other drying facilities must also be supplied. These washing facilities should not be used for any purpose other than personal hygiene. Food hygiene regulations generally contain provisions of this kind. Some insist that signs must be displayed conspicuously, instructing workers to wash their hands after using toilets.

Sanitary Facilities

Satisfactory sanitary facilities are necessary in all food premises for the use of the staff and this is usually legally enforceable. Sometimes, sanitary accommodation is also specifically required in catering establishments for customers' use. This accommodation must be adequately lighted and ventilated and kept clean and in good order.

Sleeping Places

It is usual for food hygiene legislation to prohibit the unhygienic practice of using rooms in which food is prepared, served or stored for sleeping purposes, and in some cases the legislation bans direct communication between sleeping quarters and rooms in which food is handled.

Refuse

Some regulations make it obligatory to provide adequate space, in a suitable situation, for the removal of waste from food, the separation of food fit and unfit for human consumption, and the storage of waste and unfit food before disposal. There must be no avoidable deposit or accumulation of solid or liquid wastes in a food room. Sometimes the law is more recise. It may insist, for example, that all garbage and refuse in kitchen areas should be kept in separate, leakproof, nonabsorbent containers equipped with tight-fitting covers unless otherwise protected from flies and other insects. There may be a specific obligation to maintain refuse storage areas in a clean sanitary condition. Garbage and refuse should be disposed of as often as necessary to prevent decom-position or overflow.

First-aid

The provision of first-aid equipment for food handlers is obligatory in some countries and should be generally compulsory. Storage of clothing It is usual to require lockers to be provided for the outdoor or other clothing and footwear of employees arriving for work in food handling establishments.

Water Supply

An adequate supply of clean and wholesome water is generally stipulated for all food premises. Clearly, this should be a universal requirement.

Food Protection

It is common to embody in food hygiene legislation an obligation to protect food from contamination by insects, rodents, dust, and from all other kinds of contamination. In addition, some countries specifically require the exclusion of domestic animals and birds from premises in which food is handled commercially. In some cases provision is made to avoid contamination of food by toxic materials such as insecticides, rodenticides, and other substances used for the maintenance of hygienic conditions.

The points covered are labelling of containers of toxic materials so that the contents may be easily identified, storage of these substances in locked cupboards, and their use in such a manner as not to contaminate food or endanger health. Food hygiene regulations also often contain obligations concerning the control of the temperature of food in catering establishments. These controls usually apply to perishable or potentially hazardous foods, that is, to foodstuffs that support the growth of pathogenic organisms.

The general principle embodied in this legislation is that the foodstuffs concerned should not be kept any longer than is necessary at a temperature favourable to the growth of bacteria. There is some variation in the temperature range specified but the extreme limits are 4.4°C (40°F) and 65.5°C (150°F). This range, with slight variations, is regarded as a danger zone to be avoided as far as possible.

The foods to which temperature control applies are described in different ways. In some cases the legislation refers to "perishable food" or "perishable food of animal or vegetable origin". Sometimes particular foods are referred to, such as "custards, cream fillings or similar products". In one case regulations apply to "infection and toxin prone food" which is defined as "perishable food consisting in whole or in part of milk, milk products, meat, poultry, fish, shellfish or any other ingredient capable of supporting the rapid growth of pathogenic organisms or the production of the toxins of such organisms".

In another case temperature control applies to all food "consisting of meat, fish, gravy, or imitation cream, or prepared from, or containing, any of these substances or any egg or milk". There are, however, exceptions for such articles as bread, biscuits and cakes containing egg or milk as ingredients that were introduced before baking, chocolate or sugar-confectionery, butter and other fats, cheese and uncooked bacon and ham. It is not usual to impose legal requirements for the cooking of food but some regulations provide that pork products must be thoroughly cooked to heat all parts to a minimum temperature of 65.5°C (150°F). Stuffings,-poultry, stuffed meats and stuffed poultry must be heated throughout to a minimum temperature of 74°C (165°F).

Food Handlers

It is a standard requirement of food hygiene legislation that food handlers must be clean and practise hygienic habits. They are generally required to wear clean, washable overclothing and, in some countries, to wear headgear that confines the hair. Smoking in food preparation areas is often prohibited. In some cases food handlers are specifically required to wash their hands after visits to the toilet. One set of regulations requires employees and management to be adequately informed concerning acceptable and sanitary food handling practices. They are expected to attend approved food handler training programmes when these, are available. Another set of regulations requires supervisors of food processing establish-ments, including caterers, restaurants and eating places, to complete a course in food protection. Many countries in their food hygiene legislation try to ensure that persons who become infected and who may transmit infection to food are excluded from handling food while they are in this condition or require all food handlers to undergo periodical medical examinations to determine that they are healthy and-free from infection.

For example, some regulations for restaurants require any restaurant licensee who suspects that an employee has contracted any form of communicable disease, or has become a carrier of such disease, to exclude the employee from the restaurant and notify the local health officer immediately. The health officer has to determine whether the food handler is infected or is a carrier, and laboratory examinations may be required. No persons who have a communicable disease or are carriers of such disease are permitted to work in a restaurant. Restaurants are forbidden to employ such persons or any persons suspected of being so affected.

Chest X-ray examinations and/or tuberculin tests are recommended annually for all restaurant personnel. Persons who at any time have had typhoid or paratyphoid fever may not be employed in a restaurant until it has been definitely determined that they are not carriers.

Similar regulations elsewhere provide that every food handler must be free from, and not a carrier of, a disease that could spread through the medium of food, and that food handlers must submit to such medical examinations and tests as the medical officer of health or the minister of health may require. No person who has a communicable skin disease or infection or who resides in a dwelling where there is communicable disease is permitted work as a food handler, unless-he obtains a certificate from the medical officer of health showing that he is free from infection.-The operator of a food business who knows or suspects that an employee is violating these provisions must notify the medical officer of health.

Another example of this kind of legislation is an order prohibiting persons liable to taint or contaminate food from engaging in the handling of food.-This provision applies, inter alia, to persons who:

(1) are suffering, or suspected to be suffering, from typhoid or paratyphoid fever or other Salmonella infections, dysentery, or staphylococcal infections;

(2) are carriers of the microorganisms causing these diseases;

(3) show detectable clinical symptoms of infectious hepatitis;

(4) are suffering, or suspected to, be suffering, from tuberculosis in a communicable state;

(5) are suffering, or suspected to be suffering, from a contagious disease of the skin.

This legislation also requires all persons professionally engaged in the production or distribution of foodstuffs or substances used as food and who are in direct contact with such products to undergo an annual examination to verify that they are free from tuberculosis in a communicable state.

Legislation in another country places an obligation on any person employed as a food handler and who knows that he, a member of his family, or another person who lives in the same dwelling is suffering from a dangerous disease to inform his immediate superior of the facts. The latter must immediately notify the health board. '

This legislation also provides that no person who is suffering from boils, a rash or sores, or from a dangerous disease, or from tuberculosis, leprosy or syphilis, may be permitted to engage in the handling of foodstuffs intended for sale. This also applies to carriers of disease. Where necessary, the person concerned must produce a medical certificate proving that the disease from which he is suffering is not contagious or liable to cause food poisoning.

Further, before any person is engaged, or begins to work, as a food handler he must produce a medical certificate to show that he is not suffering from a disease which, having regard to the nature of the work, could involve a danger of contagion or food poisoning through the agency of foodstuffs. The medical certificate must not have been issued more than 30 days previously. It must be kept and, on request, shown to the health and police authorities.

In yet another country, immediately a person engaged in the handling of food becomes aware he is suffering from, or is a carrier of, typhoid or paratyphoid fever or any other Salmonella infection or amoebic or bacillary dysentery or any staphylococcal infection likely to cause food poisoning, he is required to inform his employer who must immediately notify the proper officer of the local authority. Infected persons may be ordered to discontinue or to refrain from engaging in any occupation connected with the preparation and handling of food or drink for human consumption until further notice is given by the local authority that the risk of transmitting infection no longer exists. Food handlers suspected of being carriers of any of the diseases mentioned above may also be required to undergo a medical examination.

One local authority may require food workers, before employment, or at any other time, to undergo a chest X-ray examination or any other test or examination deemed necessary for the protection of the public health. Certification of such examinations may be issued by the local health authority and, when required by the authority, a copy of the certificate for each employee must be in the possession of the management of the food service establishment.

Certificates are not Valid for more than 12 Months from the Date of Issue

Certain countries require food handlers to hold a health booklet or certificate. In one of these countries workers who are to be employed in the food industry are obliged to undergo a pre-employment medical examination. This examination is necessary even for temporary employees, apprentices, and students. The workers covered by this obligation include those employed in the sale of foodstuffs and the provision of meals to the public, as well as in the distribution of snacks and refreshments.

Also in that country, food workers must undergo other preventive medical examinations at times determined by the hygiene services. In addition, they must undergo, without delay, emergency medical examinations:

(a) if they are suffering from diarrhoea, a purulent disease or a fever, or a communicable disease, or if they are suspected or suffering from a communicable disease;

(b) if a communicable disease, or a suspected case of such a disease, occurs in the work-place or at home.

These medical examinations include a detailed clinical examination, a medical history, and a microbiological examination,particularly with a view to the detection of tuberculosis and possible carriers of the pathogens of typhoid and paratyphoid and other salmonelloses and shigelloses.

Workers engaged in the food industry are obliged to hold a health booklet issued by the health community medical officer for the area in which they live or, if they work in a factory having a medical officer, by that officer. In the case of workers employed for short periods a certificate as to their medical fitness is issued in place of the booklet. The booklet (or certificate) is kept by the person in charge of the undertaking, but where workers carry out their work outside the undertaking, they are required to carry the booklet (or certificate) themselves. The person in charge of the undertaking must keep a list of the workers indicating when they were medically examined, the results of the examination, and when the next medical examination should be carried out.

At each medical examination the worker must produce his booklet or certificate. If the physician finds, or suspects, that the worker is suffering from a disease or infection he must take charge of the booklet or certificate and give the worker a written receipt in exchange. The worker is obliged to hand over the receipt, without delay, to the person in charge of the undertaking. Every food establishment must ensure that the handling of food is carried out only by workers holding a booklet or certificate; work may not be commenced before the booklet or certificate has been issued. The establishment must notify the health community medical officer in good time of every worker who is required to undergo a preventive medical examination.

Another country imposes compulsory medical examinations on certain groups of workers including those in catering establishments. In addition to the usual examinations for typhoid, paratyphoid, dysentery and other Salmonella infections, workers in catering establishments are subject to examination for infestation with intestinal parasites.

A worker who is transferred to another undertaking, establishment or installation is not subject to a prior medical examination if his health booklet shows that his initial pre-employment examination was satisfactory, that his state of health has been under systematic surveillance, and that during the previous two years, no member of his family or other person living in his household has suffered from an infectious intestinal disease or an intestinal parasitosis.

After every absence from work lasting more than one month the person concerned must be examined as if he were beginning work for the first time.

Every food handler (i.e., any person who carries out, directly or indirectly, manual activities involving foodstuffs in such a manner as to be able to affect their wholesomeness, quality or hygiene) in another country must undergo the health examinations prescribed by the health authority for the purpose of acquiring a health certificate entitling him to work in that occupation.

The certificate must be renewed annually or at more frequent intervals should the health authority consider it necessary. Any person who, while not holding the appropriate health certificate, engages in food handling and the proprietor of the establishment where he works are both liable to penalties.

Another example which may be mentioned is an ordinance concerning hygiene in restaurants and catering establishments. Every worker must undergo a pre-employment medical examination to determine his suitability for the work he will be engaged in.

The nature of the examinations required, the classes of workers subject to such examinations, and the frequency of, and procedure for, the medical examinations are prescribed.

A recent piece of legislation on food hygiene states that only persons who provide a medical certificate indicating that they are free from diseases or infections that could be transmitted to other persons by means of foodstuffs may commence work involving the handling of foodstuffs other than pre-packed foodstuffs, or work directly related to hygiene in the handling of foodstuffs, in various types of premises including restaurants and other catering establishments.

This legislation also requires persons working in the types of premises to which it applies to undergo at least once a year a medical examination or other health checkup as officially prescribed.

Physicians or veterinarians who have good grounds for believing that food poisoning has been caused by a foodstuff contaminated by pathogenic bacteria or extraneous substances during food handling operations are required to notify the local medical officer.

Basic Principles of Food Hygiene Legislation

General

An essential part of the health services of every country must be to ensure that food is safe, wholesome and not injurious to health, and that proper hygienic standards are observed at all stages from production or manufacture until it reaches the consumer. The requirement that food must not be injurious to the consumer means that food traders must be under a legal obligation not to sell food which they know to be unfit for human consumption or could with reasonable diligence ascertain to be unfit. They must also have a duty to take due care to ensure that adequate standards of food hygiene are observed at all times in the conduct of their business.

It is not sufficient merely to place legal obligations on food traders. There must also be official agencies charged with the duty of exercising close surveillance of all types of food premises, including catering establishments. Many environmental factors can influence the maintenance of satisfactory standards of food hygiene and, as indicated on p. 3, progress will be made more easily if a food hygiene programme is developed within a progressive environmental health and personal health programme.

Where food hygiene is the responsibility of the agency responsible for general environmental health there should be no problems in securing coordination of effort. Where separate agencies are responsible it is important that there should be close liaison between them. Agencies with the duty of controlling food hygiene should not be regarded merely as law enforcement bodies, though they should possess adequate legal powers to establish and maintain the required standards. Their officers need to have authority to enter food premises during business hours without warning in order to carry out the necessary inspections and investigations, and it should be an offence to obstruct officers who are carrying out these duties.

Effective legislation is necessary to secure the proper observance of hygienic practices. The basic legal requirements should be embodied in national statutes but the more detailed and technical requirements should be contained in regulations which can be readily and quickly changed to keep pace with scientific and technical developments. There is a need also for codes of practice which, while not legally enforceable, give guidance to food traders and the officials responsible for inspection and control of food premises on equipment, materials and practices for maintaining the desired standards.

Although the basic legal requirements embodied in national statutes and the more detailed and technical requirements contained in regulations will cover very many points it is not advisable that legal provisions should be too rigid or too detailed. Often, the more satisfactory approach is to prescribe the objective to be achieved and to allow some flexibility in the choice of methods to be used to achieve that objective. In any case, it is impracticable for laws to be drafted in sufficient detail to take account of all possible circumstances.

Basic Legal Requirements

The principal points that should be included in the basic legal requirements are as follows :

(a) the authorities responsible for the enforcement of food hygiene legislation should be specified;

(b) it should be an offence to sell food that is diseased, unsound, or unwholesome;

(c) recognized officers of the responsible authorities should be authorized to inspect food intended for sale, to take samples, and to seize and condemn food that appears to be unfit for human consumption;

(d) recognized officers should have power of entry to all food premises;

(e) all food premises of prescribed types should be required to be registered by the responsible authorities before being opened for business and these authorities should be empowered to revoke registration where continued use of the premises constitutes a potential danger to health; particular types of premises would be registered under regulations made in accordance with point (f) below;

(f) authority should be given to the appropriate minister to make regulations containing specific and detailed legal requirements with respect to food hygiene in particular types of premises and different food handling activities; compliance with the appropriate regulations would be a condition for the registration of premises;

(g) physicians who become aware, or suspect, that patients they are attending are suffering from food poisoning should be required to notify the local public health authority;

(h) authority should be given to recognized officers to prohibit the use or removal from the premises of food which they believe is likely to cause food poisoning until investigations have shown whether it may safely be used for human consumption;

(i) persons engaged in the handling of food should be required to notify their employers if they become aware that they are suffering from any condition that might lead to the infection of foods and employers should be under an obligation to inform the local medical officer.

Model Food Hygiene Regulations for Catering Establishments

The basic legal requirements for the control of food hygiene which should be embodied in a national statute have been set out in the preceding section.

As indicated in point (f), detailed legal requirements should be stated in regulations or secondary legislation made under the authority given by the statute. It should be noted that in these model regulations expressions such as “adequate”, “suitable”, and “sufficient” are frequently used.

This is because a legal document of this kind must have sufficient flexibility to meet the varying circumstances found in different types of establishment and in different parts of the world. The interpretation of these terms is left to the discretion of local control authorities.

The following Model Code details those matters which it is suggested should be included in the food hygiene regulations for catering establishments.

Registration

(a) No premises shall be used as catering premises unless they are registered for that purpose by the local authority. Any person who uses unregistered premises shall be guilty of an offence.

(b) Application for registration must be made before the premises are opened and the application must be refused if the premises do not comply in all respects with food hygiene regulations for catering premises.

(c) Registration may be revoked if at any time serious breaches of the food hygiene regulations occur.

Where a licensing system is preferred, points (e) and (f) should be amended accordingly.

Where a system of licensing is preferred this section should be amended accordingly.

Construction and Maintenance of Premises

(a) No catering business shall be carried on in any premises in which food is exposed to the risk of contamination.

(b) The premises shall be of adequate size, sound construction and in good repair. Areas used for the handling of food shall be completely separated from any part of the premises used as living quarters.

(c) The internal walls, floors, windows, ceilings, woodwork and all other parts of every room shall be so constructed and maintained in such good order, repair and condition as to enable them to be cleaned effectively and to prevent, as far as practicable, infestation by rats, mice, insects and birds.

(d) The premises must be kept free from rodent and insect pests, and birds, and dogs, cats and other domestic animals must be excluded from areas where food is processed or stored. All rodenticides, insecticides and other toxic substances must be clearly labelled and, when not in use, be kept in separate, locked cupboards.

(e) The premises must be kept in a clean condition and no water must be deposited or allowed to accumulate in rooms in which food is prepared, stored or served, except in so far as this may be unavoidable for the proper conduct of the business.

(f) All rooms in which food is stored, prepared or served, or in which utensils are washed, and all sanitary conveniences shall be well ventilated and well lighted.

(g) No room in which food is stored, prepared or served may be used as a sleeping place or communicate directly with a sleeping place or sanitary convenience.

Water Supply

(a) An ample supply of water of drinking quality must be provided at all premises; where possible this should be from the public supply main but if from any other source it must be approved by the health authority and checked for bacteriological and chemical quality at intervals.

(b) An adequate supply of hot and cold water or of hot water at a suitable controlled temperature shall be provided in all areas where food is prepared and where equipment and utensils are washed.

(c) Water used for making ice shall be of drinking quality.

Drainage

(a) All premises shall be provided with an adequate drainage system which, where practicable, shall be connected to the public sewerage system.

(b) Where connexion to the public sewerage system is not practicable the method of sewage disposal must be approved by the health authority.

Personal Washing Facilities

(a) An adequate number of suitable wash-hand basins shall be provided for the use of all persons engaged in the handling of food. These basins must be conveniently accessible to these persons and in any case must be located in or immediately adjacent to all sanitary conveniences.

In all new premises, and where practicable in existing premises, hand-washing facilities must be provided within or immediately adjacent to food preparation areas.

All wash-hand basins shall be properly connected to the drainage system and be provided with an adequate supply of hot and cold water or of hot water at a suitable controlled temperature. At each basin shall be provided an adequate supply of soap or other suitable detergent, nail brushes and clean towels or other suitable drying facilities.

(b) Facilities for personal washing must not be used for any other purpose.

Sanitary Facilities

(a) An adequate number of conveniently located sanitary conveniences shall be provided for all employees. Separate conveniences shall be provided for each sex. An adequate supply of toilet paper must be available at all times.

(b) In or near every sanitary convenience a clearly legible notice must be displayed in a prominent position requiring users to wash their hands after using the convenience.

Storage of Clothing

Adequate facilities shall be provided for the storage of employees' clothing and personal belongings not being worn during working hours and such clothing and belongings shall not be kept on the premises other than in the accommodation provided.

First-aid Materials

(a) A sufficient supply of suitable bandages, dressings (including waterproof dressings) and antiseptics shall be provided for the first-aid treatment of persons engaged in the handling of food.

(b) This material must be kept in suitable locations readily accessible to employees and clearly labelled.

Storage of Refuse

(a) A sufficient number of watertight containers of durable and nonabsorbent material with tight-fitting or self-closing lids shall be provided for the storage of food waste and other refuse.

(b) Adequate space shall be provided for the temporary storage of waste; this space must not be in any room where food is stored, prepared or served.

Food

(a) Food, while being stored, prepared, displayed or served, shall be protected from dust, flies, rodents and other pests, and other contamination and all persons engaged in the handling of food shall take all reasonable steps to protect the food from risk of contamination.

(b) Raw meat, poultry, fish, shellfish, milk or uncooked products containing those foodstuffs shall not be permitted to come into contact with any cooked foods.

Any persons handling any of these raw or uncooked foodstuffs must, before handling any cooked food, thoroughly wash their hands and carefully cleanse all surfaces, utensils and equipment with which the foodstuffs may have come into contact.

(c) All perishable food consisting in whole or in part of milk or milk products, eggs, meat, poultry, fish, shellfish or other ingredients capable of supporting the rapid growth of pathogenic organisms shall be kept at a temperature that is not within the range from 4.5°C (40°F) to 65.5CC (150°F), except during necessary periods of preparation and service.

(d) Raw, unprocessed fruits and vegetables shall be thoroughly washed in clean drinking-water before use.

(e) All ice used for cooling drinks or food by direct contact shall be made from water of drinking quality.

Equipment

(a) All articles of equipment with which food comes into contact, or may come into contact, must be kept clean, be so constructed, placed, of such materials, and kept in such good order, repair and condition as to:

(i) enable them to be thoroughly cleaned;

(ii) prevent, so far as is reasonably practicable, any matter being absorbed by them; and

(iii) prevent, so far as is reasonably practicable, any risk of contamination of the food.

(b) An adequate number of sinks or other suitable facilities for the washing of food and equipment when necessary shall be provided. These sinks or other facilities shall be provided with an adequate supply of hot and cold water or of hot water at a suitable controlled temperature, or with cold water only where the sink is used only for washing fish, fruit or vegetables. These facilities must not be used for any other purpose.

(c) Equipment shall be so installed as to facilitate its cleaning and the cleaning of adjacent areas.

Personnel

(a) Every person handling or coming into contact with food, or with any equipment or utensil used in its preparation, processing or service shall:

   (i) keep as clean as may be reasonably practicable all parts of his body that are liable to come into contact with the food and, in particular, he shall wash his hands thoroughly before commencing work and after using a sanitary convenience;

   (ii) wear clean, washable outer garments and headgear that confines his hair;

   (iii) keep any open cut or abrasion on any exposed part of his body covered with a suitable waterproof dressing; and

   (iv) refrain from spitting and using tobacco in any form while engaged in the handling, preparation or service of food, while handling utensils or equipment, and while he is in any room containing open food.

(b) No person who is affected with any disease in a communicable form may work in any catering premises and the proprietor of any catering establishment shall not permit any such person to work in the establishment. Immediately a person engaged in the handling of food becomes aware, or has reason to suspect, that he is suffering from, or is a carrier of, typhoid, paratyphoid or any other Salmonella infection or amoebic or bacillary dysentery or any staphylococcal infection likely to cause food poisoning he shall immediately inform the person operating the catering premises in which he is employed and that person shall immediately notify the medical officer of the health authority.

Food & Beverage Department-Restaurant Division

Out of the three divisions that the Food & Beverage Department holds, the Restaurant Division is the most popular when it comes to money. Its structure has changed recently, but it's fairly simple and everyone knows who to report to. Described in a simple way, the structure goes like this:

Restaurant Operations Manager

He or she is a three stripe officer and is in charge with the entire restaurant operation. The Windjammer Manager, Portofino Manager and Assistant Dining Room Manager report directly to him or her and, of course, the position is got its own admin now.

The contract length is 4 months with 2 months off and, aside the thousands of dollars salary, he or she is got monthly bonuses from china and linen savings, wine gratuity and the big, fat bonus at the end of the fiscal year.

Windjammer Manager

The Windjammer Manager is a two and a half stripes officer and is in charge with Windjammer Buffet, Johnny Rockets (fast food outlet) and Café Promenade (coffee shop with snacks). Directly to him reports the following positions: Johnny Rockets Supervisor and Windjammer Supervisor.

Just like the Restaurant Operations Manager, the contract length is also 4 months on with 2 months off, and a big, fat bonus included.

Portofino Manager

Portofino Manager is in charge with the Italian specialty restaurant. Is a two stripes officer with a 6 months contract and sharing cabin. From 2 stripes below, no officer is entitled to the yearly bonus, but they got part of the cover charge. The Portofino waiters report directly to him.

Assistant Dining Room Managr

This position was introduced recently. It's a 2 stripe position with a contract length of 6 months and sharing cabin. Their salary is based on tips, but don't be fooled by it, they make a lot of money. There are 3 Assistant Dining Room Managers, each in charge with one Dining Room deck (our Dining Room has 3 decks/levels). They are responsible with the booking on their respective decks and the Headwaiters working in that deck report directly to them.

Restaurant Operation Admin

The admin reports directly to the Restaurant Operations Manager and basically does everything.

He has 2 office boys/girls helping in the morning with paperwork. Is a one stripe officer with 6 months contract and sharing cabin.

Head Waiter

These guys are making more money than a 3 stripe officer because their salary is based on tips.

They are one stripe officers, with 6 months contract length and a sharing cabin. Each of them has a station assigned (there are 3 stations on each Dining Room deck) and the waiters and assistant waiters working there report to him/her.

Johnny Rockets Supervisor

He or she is in charge with Café Promenade and Johnny Rockets and reports to the Windjammer Manager. This is a one stripe officer with a 6 months contract and sharing cabin. They get a fix salary with no gratuities. The Café Attendants and Johnny Rockets Attendants report to him/her.

Windjammer Supervisor

There are 2 Windjammer Supervisors onboard, both reporting to the Windjammer Manager. They are in charge with supervising the Windjammer operation and the Assistant Waiters working there are reporting to them. It's a one stripe position, with fix salary, 6 months contract and sharing cabin.

Room Service Supervisor

The Room Service Supervisor reports directly to the Restaurant Operations Manager and is in charge with Room Service operation and Attendants working there. Is a one stripe position, with 6 months contract, fix salary and sharing cabin.

Room Service Operator

The position is pretty much self-explanatory. They are part of the crew with a 6 months contract, fix salary and sharing cabin. The Room Service Operator reports to Room Service Supervisor.

Mess Supervisor

The Mess Supervisor is in charge with both officer and crew messes and directly to him/her are reporting the Mess Attendants. Mess Supervisor is a one stripe position with a 6 months contract, fix salary and sharing cabin. He or she reports to Restaurant Operations Manager.

Room Service Attendant

The above 5 positions are similar; the only difference is the place they work in. Except for the Johnny Rockets who get gratuities from the cover charges in the fast food outlet they work in, the rest have a fix salary and they also get gratuities from the beverages they sell (not too much though). They all have an eight months contract and sharing cabin.

Assistant Waiter

This is a much better position that the ones above because they have a salary based on tips. An average assistant waiter salary can be around $2000 or more a month. Also, the contract length is 6 months and they only work in the Dining Room or Windjammer during the dinner time and in Windjammer (during the day time). During the diner time, they work 4 weeks in the Dining Room and one week in Windjammer. They get gratuities in the Dining Room, but in Windjammer they get only the gratuities from the beverages they sell. Might not make sense, but this is how the system is set up.

Waiter

Waiter position is a good position as well due to the amount of money they make and the 6 months contract length. An average waiter makes around $3000 and, during the diner time, works only in Dining Room. As a first-timer, this structure might look confusing. Indeed there are a lot of positions in the Restaurant Division, but this is what makes it easier because everyone knows exactly what they are suppose to do.

Food and Beverage Service Occupations

Significant Points

- Most jobs are part time and many opportunities exist for young people—nearly 2 out of 3 food counter and fountain workers are 16-19 years old.
- Job openings are expected to be abundant through 2008, reflecting substantial turnover.
- Tips comprise a major portion of earnings; consequently, keen competition is expected for bartender, waiter and waitress, and other jobs in popular restaurants and fine dining establishments where potential earnings from tips are greatest.

Nature of the Work

Whether they work in small, informal diners or large, elegant restaurants, all food and beverage service workers aim to help customers have a positive dining experience in their establishments. These workers are responsible for greeting customers, taking food and drink orders, serving food, cleaning up after patrons, and preparing tables and dining areas. All of these duties require a high quality of services customers will return.

The largest group of these workers, *waiters* and *waitresses,* take customers' orders, serve food and beverages, prepare itemized checks, and sometimes accept payments. Their specific duties vary considerably, depending on the establishment where they work. In coffee shops, they are expected to provide fast and efficient, yet courteous service.

In fine restaurants, where gourmet meals are accompanied by attentive formal service, waiters and waitresses serve meals at a more leisurely pace and offer more personal service to patrons. For example, servers may recommend a certain wine as a complement to a particular entree, explain how various items on the menu are prepared, or complete preparations on a salad or other special dishes at table side. Additionally, waiters and waitresses may check the identification of patrons to ensure they meet the minimum age requirement for the purchase of alcohol and tobacco products.

Depending on the type of restaurant, waiters and waitresses may perform additional duties usually associated with other food and beverage service occupations. These tasks may include escorting guests to tables, serving customers seated at counters, setting up and clearing tables, or operating a cash register. However, formal restaurants frequently hire other staff to perform these duties, allowing their waiters and waitresses to concentrate on customer service.

Bartenders fill drink orders that waiters and waitresses take from customers. They prepare standard mixed drinks and, occasionally, are asked to mix drinks to suit a customer's taste. Most bartenders know dozens of drink recipes and are able to mix drinks accurately, quickly, and without waste, even during the busiest periods.

Besides mixing and serving drinks, bartenders collect payment, operate the cash register, clean up after customers leave, and often serve food to customers seated at the bar. Bartenders also check identification of customers seated at the bar, to ensure they meet the minimum age requirement for the purchase of alcohol and tobacco products. Bartenders usually are responsible for ordering and maintaining an inventory of liquor, mixes, and other bar supplies. They often form attractive displays out of bottles and glassware and wash the glassware and utensils after each use.

The majority of bartenders who work in eating and drinking establishments directly serve and interact with patrons. Because customers typically frequent drinking establishments for the friendly atmosphere, most bartenders must be friendly and helpful with customers.

Bartenders at service bars, on the other hand, have little contact with customers because they work in small bars in restaurants, hotels, and clubs where only waiters and waitresses serve drinks. Some establishments, especially larger ones, use automatic equipment to mix drinks of varying complexity at the push of a button. Even in these establishments, however, bartenders still must be efficient and knowledgeable in case the device malfunctions or a customer requests a drink not handled by the equipment.

Hosts and *hostesses* try to create a good impression of a restaurant by warmly welcoming guests. Because hosts and hostesses are restaurants' personal representatives, they try to insure that service is prompt and courteous and that the meal meets expectations. They may courteously direct patrons to where coats and other personal items may be left and indicate where patrons can wait until their table is ready. Hosts and hostesses assign guests to tables suitable for the size of their group, escort patrons to their seats, and provide menus. They also schedule dining reservations, arrange parties, and organize any special services that are required. In some restaurants, they also act as cashiers.

Dining room attendants and *bartender helpers* assist waiters, waitresses, and bartenders by cleaning tables, removing dirty dishes, and keeping serving areas stocked with supplies. They replenish the supply of clean linens, dishes, silverware, and glasses in the dining room and keep the bar stocked with glasses, liquor, ice, and drink garnishes. Bartender helpers also keep bar equipment clean and wash glasses. Dining room attendants set tables with clean tablecloths, napkins, silverware, glasses, and dishes and serve ice water, rolls, and butter. At the conclusion of meals, they remove dirty dishes and soiled linens from tables. Cafeteria attendants stock serving tables with food, trays, dishes, and silverware and may carry trays to dining tables for patrons.

Counter attendants take orders and serve food at counters. In cafeterias, they serve food displayed on counters and steam tables, carve meat, dish out vegetables, ladle sauces and soups, and fill beverage glasses. In lunchrooms and coffee shops, counter attendants take orders from customers seated at the counter, transmit orders to the kitchen, and pick up and serve food. They also fill cups with coffee, soda, and other beverages and prepare fountain specialties, such as milkshakes and ice cream sundaes. Counter attendants prepare some short-order items, such as sandwiches and salads, and wrap or place orders in containers for carry out. They also clean counters, write itemized checks, and sometimes accept payment.

Fast-food workers take orders from customers at counters or drive-through windows at fast-food restaurants. They pick up the ordered beverage and food items, serve them to a customer, and accept payment. Many fast-food workers also cook and package food, make coffee, and fill beverage cups using drink-dispensing machines.

Working Conditions

Food and beverage service workers are on their feet most of the time and often carry heavy trays of food, dishes, and glassware. During busy dining periods, they are under pressure to serve customers quickly and efficiently. The work is relatively safe, but care must be taken to avoid slips, falls, and burns.

Part-time work is more common among food and beverage service workers than among workers in almost any other occupation. Those on part-time schedules include half of all waiters and waitresses, and 6 out of 10 food counter and fountain workers, compared to 1 out of 6 workers throughout the economy. Slightly more than half of all bartenders work full-time with 35 percent working part-time and the remainder working a variable schedule. The wide range in dining hours creates work opportunities attractive to homemakers, students, and other individuals seeking supplemental income. In fact, nearly 2 out of 3 food counter and fountain workers are between 16 and 19 years old. Many food and beverage service workers work evenings, weekends, and holidays. Some work split shifts—that is, they work for several hours during the middle of the day, take a few hours off in the afternoon, and then return to their jobs for evening hours.

Employment

Food and beverage service workers held over 5.4 million jobs in 1998. Waiters and waitresses held about 2,019,000 of these jobs; counter attendants and fast-food workers, 2,025,000; dining room and cafeteria attendants and bartender helpers, 405,000; bartenders, 404,000; hosts and hostesses, 297,000; and all other food preparation and service workers, 280,000.

Restaurants, coffee shops, bars, and other retail eating and drinking places employed the overwhelming majority of food and beverage service workers. Others worked in hotels and other lodging places, bowling alleys, casinos, country clubs, and other membership organizations. Jobs are located throughout the country but are typically plentiful in large cities and tourist areas. Vacation resorts offer seasonal employment, and some workers alternate between summer and winter resorts, instead of remaining in one area the entire year.

Training, Other Qualifications, and Advancement

There are no specific educational requirements for food and beverage service jobs. Although many employers prefer to hire high school graduates for waiter and waitress, bartender, and host and hostess positions, completion of high school is usually not required for fast-food workers, counter attendants, and dining room attendants and bartender helpers. For many people, a job as a food and beverage service worker serves as a source of immediate income, rather than a career. Many entrants to these jobs are in their late teens or early twenties and have a high school education or less. Usually, they have little or no work experience. Many are full-time students or homemakers. Food and beverage service jobs are a major source of part-time employment for high school and college students.

Because maintaining a restaurant's image is important to its success, employers emphasize personal qualities. Food and beverage service workers are in close contact with the public, so these workers should be well-spoken and have a neat, clean appearance. They should enjoy dealing with all kinds of people and possess a pleasant disposition.

Waiters and waitresses need a good memory to avoid confusing customers' orders and to recall faces, names, and preferences of frequent patrons. These workers should also be good at arithmetic so they can total bills without the assistance of a calculator or cash register if necessary. In restaurants specializing in foreign foods, knowledge of a foreign language is helpful. Prior experience waiting on tables is preferred by restaurants and hotels that have rigid table service standards. Jobs at these establishments often have higher earnings, but they may also have higher educational requirements than less demanding establishments.

Usually, bartenders must be at least 21 years of age, but employers prefer to hire people who are 25 or older. Bartenders should be familiar with State and local laws concerning the sale of alcoholic beverages. Most food and beverage service workers pick up their skills on the job by observing and working with more experienced workers. Some employers, particularly those in fast-food restaurants, use self-instruction programs with audiovisual presentations and instructional booklets to teach new employees food preparation and service skills. Some public and private vocational schools, restaurant associations, and large restaurant chains provide classroom training in a generalized food service curriculum.

Some bartenders acquire their skills by attending a bartending or vocational and technical school. These programs often include instruction on State and local laws and regulations, cocktail recipes, attire and conduct, and stocking a bar. Some of these schools help their graduates find jobs.

Due to the relatively small size of most food-serving establishments, opportunities for promotion are limited. After gaining some experience, some dining room and cafeteria attendants and bartender helpers are able to advance to waiter, waitress, or bartender jobs. For waiters, waitresses, and bartenders, advancement usually is limited to finding a job in a more expensive restaurant or bar where prospects for tip earnings are better. A few bartenders open their own businesses. Some hosts and hostesses and waiters and waitresses advance to supervisory jobs, such as maitre d'hotel, dining room supervisor, or restaurant manager. In larger restaurant chains, food and beverage service workers who excel at their work are often invited to enter the company's formal management training program.

Job Outlook

Job openings are expected to be abundant for food and beverage service workers. Employment of food and beverage service occupations is expected to grow about as fast as the average for all occupations through 2008, stemming from increases in population, personal incomes, and leisure time. While employment growth will produce many new jobs, the overwhelming majority of openings will arise from the need to replace the high proportion of workers who leave this occupation each year.

There is substantial movement into and out of the occupation because education and training requirements are minimal, and the predominance of part-time jobs is attractive to people seeking a short-term source of income rather than a career. However, keen competition is expected for bartender, waiter and waitress, and other food and beverage service jobs in popular restaurants and fine dining establishments, where potential earnings from tips are greatest.

Projected employment growth will vary by type of food and beverage service job. Growth in the number of families and the more affluent, 55-and-older population will result in more restaurants that offer table service and more varied menus—requiring waiters and waitresses and hosts and hostesses. Employment of fast-food workers also is expected to increase in response to the continuing fast-paced lifestyle of many Americans and the addition of healthier foods at many of these restaurants. However, little change is expected in the employment of dining room attendants, as waiters and waitresses increasingly assume their duties. Employment of bartenders is expected to decline as drinking of alcoholic beverages outside the home—particularly cocktails—continues to drop.

Earnings

Food and beverage service workers derive their earnings from a combination of hourly wages and customer tips. Earnings vary greatly, depending on the type of job and establishment. For example, fast-food workers and hosts and hostesses usually do not receive tips, so their wage rates may be higher than those of waiters and waitresses and bartenders, who may earn more from tips than from wages. In some restaurants, these workers contribute a portion of their tips to a tip pool, which is distributed among the establishment's other food and beverage service workers and kitchen staff.

Tip pools allow workers who normally do not receive tips, such as dining room attendants, to share in the rewards of a well-served meal. In 1998, median hourly earnings (not including tips) of full-time waiters and waitresses were $5.85. The middle 50 percent earned between $5.58 and $6.32; the top 10 percent earned at least $7.83. For most waiters and waitresses, higher earnings are primarily the result of receiving more in tips rather than higher hourly wages. Tips usually average between 10 and 20 percent of guests' checks, so waiters and waitresses working in busy, expensive restaurants earn the most.

Full-time bartenders had medianhourly earnings (not including tips) of $6.25 in 1998. The middle 50 percent earned from $5.72 and $7.71; the top 10 percent earned at least $9.19 an hour. Like waiters and waitresses, bartenders employed in public bars may receive more than half of their earnings as tips. Service bartenders are often paid higher hourly wages to offset their lower tip earnings.

Median weekly hourly earnings (not including tips) of full-time dining room attendants and bartender helpers were $6.03 in 1998. The middle 50 percent earned between $5.67 and $7.11; the top 10 percent earned over $8.49 an hour. Most received over half of their earnings as wages; the rest of their income was a share of the proceeds from tip pools.

Full-time counter attendants and fast-food workers, except cooks, had median hourly earnings (not including tips) of $6.06 in 1998. The middle 50 percent earned between $5.67 and $7.14, while the highest 10 percent earned over $8.45 a hour. Although some counter attendants receive part of their earnings as tips, fast-food workers usually do not.

In establishments covered by Federal law, most workers beginning at the minimum wage earned $5.15 an hour in 1998. However, various minimum wage exceptions apply under specific circumstances to disabled workers, full-time students, youth under age 20 in their first 90 days of employment, tipped employees, and student-learners. Employers are also permitted to deduct from wages the cost, or fair value, of any meals or lodging provided. However, many employers provide free meals and furnish uniforms. Food and beverage service workers who work full time often receive typical benefits, while part-time workers usually do not.

In some large restaurants and hotels, food and beverage service workers belong to unions—principally the Hotel Employees and Restaurant Employees International Union and the Service Employees International Union.

Perspectives on Productivity

Most studies of productivity are drawn from manufacturing industries, however, services require an approach that recognises their peculiar characteristics. In the provision of services the processes of production and consumption are often simultaneous or can overlap-generating a specific set of challenges (Johnston & Jones, 2004; Mullins, 1998). Customers are active participants in many service encounters, and therefore an important influence on productivity (Martin et al., 2001). Quality is so closely entwined with more measurable outcomes in service provision that it becomes very difficult to isolate any one influence on productivity. The intangibility of service, and therefore the importance of psychological outcomes (e.g. comfort and pleasure) in the process of quality creation, represent major challenges in measuring and understanding service sector productivity in general, and the F&B sector in particular.

This section of the report reviews the perspectives held b both employers and employees on productivity in their NZ F&B work-places and the factors that influence it. The thematic areas addressed below reflect the core themes raised by the 20 interviewees and the focus groups employee participants. Survey responses are also reviewed.

What is Productivity and how is it Measured?

The issue of how to measure productivity in the service sector is a critical one for this study. Gummesson (1998) suggests that measurements of service productivity can be 'ambiguous and inadequate', resulting in unhelpful comparisons between industries.

Food and beverage service providers often focus on reducing labour costs while maintaining sales (labour generally comprises around a third of hospitality costs, and food or beverage products, another third) (Reynolds, 2004). Concentrating on the reduction of labour costs as a percentage of sales may achieve short term productivity targets but can also jeopardise long term viability due to the erosion of service standards. Poor service affects customer satisfaction, which in turn influences sales and productivity, thereby creating a cycle of poor productivity.

Some commentators argue that a primary focus in F&B service productivity is quality. Parasuraman (2002) argues that how a company allocates service inputs will directly affect productivity from the customer's perspective. This view is upheld by Bates, Bates and Johnston (2003) who found that providing better service, while labour intensive, yields higher long-term profits per employee. They concluded that better service providers have significantly better return on equity and return on total assets than their counterparts with poorer service levels.

Measuring productivity in hospitality is acknowledged to be particularly difficult (Atkinson & Brown, 2001). David, Grabski and Kasavana (1996) found nine different measurements of productivity alone in their survey of hotel chain bench-marking. Most food service contract companies use sales per labour hour to express productivity (Clark, 1997a), but according to Reynolds, do not measure the cost of food sales, which he views as a 'good measure of labour productivity, but not of operational performance' (1998).

The heavy use of part-time and seasonal labour by F&B enterprises adds a further level of complexity to measuring productivity in the sector. The situation in NZ mirrors the international context quite closely. It is clear that several managers and employees do not have a good understanding of productivity and what it means in the context of their business.

As the interviews and focus groups progressed, however, it became obvious that factors related to, and influenced by, productivity were very much in people's minds-its just that productivity was not always the catchall term used to help think about them.

Both the interviews and survey show that managers view customer satisfaction and loyalty rate as the most important factors to use in assessing productivity (74% and 50% of survey respondents), whereas financial measures (meeting budget, costs and balancing, and revenue/ sales per person) are 'most important' indicators for under a third of respondents.

***Table** : Indicators used to assess productivity*

	Most important (%)	Least important (%)
Customer Satisfaction	74	11
Return visitation / loyalty	50	12
Volume	30	30
Meeting Budget	27	27
Cost and Balancing	24	45
Revenue / Sales per person	18	41
Number of table turns	13	50
Covers Per Staff	0	67

Larger F&B enterprises (especially those housed in hotels) have fairly sophisticated financial targets on which productivity is measured.

> *"[Productivity is] the amount of product that gets prepared and cooked and sold in relation to the number of staff I have... how many man hours for our front of house staff measured against revenue for a particular night".*

Labour costs and food costs are the most critical and closely reviewed areas underpinning productivity. Fast food/takeaway outlets were found to be acutely aware of the various 'actual' versus 'desired' ratios they are working to.

Nevertheless for most managers (regardless of size or type of enterprise) productivity is defined by customer satisfaction. Favourable feedback from customers through comment cards, tips, and unsolicited remarks are viewed as vital indicators of productivity.

> *"Productivity is about how customers feel. The quality of the feedback we get from customers in terms of positive comments and tips".*
>
> *"Communication between customers and staff-observation of their relationships tells me about the level of quality, and therefore, productivity".*

Measurement of customer satisfaction ranged from sophisticated feedback forms (larger accommodation based operations) through to the simple observation of client behaviour, responses and tipping (smaller operations). In many cases smaller operators are going by their "gut-feelings" and the number of regulars that return with friends and new clientele are a critical 'measure' of productivity.

Productivity is also correlated with staff being able to meet or exceed expectations in a team setting.

> *"Defining and measuring productivity must be seen in terms of teamwork. Individual performance leads to team work. And if that works, it flows smoothly. And if it doesn't, we can then isolate with whom the problems lie".*

Contented staff and management capability are the two factors management deem to have the most positive influence on productivity, followed by 'internal communication', 'training' and 'employee retention'. The survey results were very much reflecting the information gathered from the interviews-with managers across size and type of enterprise all stressing the importance of staff happiness, communication, and their own ability to manage these situations effectively.

***Table** : Factors impacting on productivity (% of respondents)*

	Negative influence (%)	*Positive influence (%)*	*Mean**
Contented staff	0	85	4.9
Management capability	0	83	4.8
Internal communication	0	62	4.6
Training	2	70	4.5
Employee retention	2	62	4.5
Technology	0	40	4.3
Labour outsourcing	6	9	3.1
Seasonality	15	19	3.1
Employee turnover	27	4	2.2

* *on a scale from negative influence (1) to positive influence (5)*

When asked to comment on what they think would most help improve productivity in their business, managers had a wide range of suggestions which were very similar with those raised in the interviews. Staff attitude to work and ability to work autonomously was the most common factor suggested (22.5%). Staff retention/experience, staff skills/qualifications and training were each commented on by 10% of respondents. Financial considerations at a macro level (e.g. taxation) were raised by 15% of respondents.

Communication and networking was also considered important by 15% of respondents. Supply related factors were raised in a further 20% of cases. Over half of national survey respondents (52.3%) said that the primary barriers to improving their productivity are finding and retaining good staff, lack of qualifications/knowledge/skills, and poor employee attitudes to work.

Difficulties with the property/product upon which the business is based were also a cause for some concern. Interestingly very few workplaces featured productivity incentives with the exception of some of the larger enterprises. Some comments illustrated the frustrations experienced by some managers:

> *"In my experience, most people will do the least amount of work required until management are forced to make them be more accountable..." The majority of employees are not given individual productivity-related targets to*

> *meet, although non-monetary targets such as quality service are often emphasized. Employees in all focus groups said that general head count, 'table turns', smoothness of daily operations and how much they 'up sold' in a day are good indicators of productivity. Managers were often perceived as "discussing numbers behind closed doors" and targets are not communicated to employees, or the targets do not mean much to the employees.*

Those working in larger organizations are more likely to have financial targets to be met each day and/or each week; and they are also encouraged to up-sell by offering customers drinks with their meals and desserts in order to meet targets. In larger enterprises, there are individual staff incentives for achieving productivity related targets, such as prizes and rewards:

> *"If we sell an extra breakfast we get one point, if we upgrade a room we get two points. When we reach 400 points we can stay for one night in the hotel free. For 20 points we may be able to park our car for one day free".*

Employees said that being efficient at what one does is an indicator of productivity. This includes taking the orders correctly, preparing and serving food quickly, reducing food wastage, and ensuring there are enough resources (i.e. clean glasses and cutlery) to provide to the customers. Participants across all focus groups said good customer service is an essential indicator of productivity. For them good customer service includes *"being really nice to the guests"*, *"being fast"* but *"not too fast, which can be seen as abrupt"*, having strong abilities for multitasking and *"doing little jobs, which usually don't belong to anyone in particular"*.

The Central Role of Management

The research presented above reveals quite clearly the central importance of effective management in achieving higher productivity in the F&B sector. In an industry as labour-intensive as F&B, people, and their skills, are central to all attempts to enhance productivity (D'Annunzio-Green, Maxwell & Watson, 2000). Research has shown the positive relationship between effective human resource management and superior organisational performance in many different sectors and nations (Drucker, 1994; Huselid, 1995; Wright and Haggerty, 2005; Neal, West & Patterson, 2005). Such an approach is consistent with current views on the positive role that empowerment and high-involvement work practices can play on productivity improvement.

Such approaches shift the focus from scientific management practices (i.e. task simplification and specialisation) to one that relies more on the intellectual contribution of employees, and emphasises the responsibility for productivity lying at the employee, rather than solely at the management, level. Gavin and Mason (2004) recommend that organisations be designed and managed to promote worker satisfaction, and suggest that healthy and happy employees are more productive.

In other research service strategies that took account of the need for employee development and well-being were found to have a positive effect on employee satisfaction and productivity (Goldstein, 2003).

As the Department of Labour (2004) notes in its discussion of productivity drivers, there is a:

> *"need to make sure the activities that create value within a firm are aligned with each other and with the overall business strategy, and that they are functioning effectively."*

This can be aided by employee participation.

> *"It is critical that employees at all levels of a firm have an opportunity to contribute to work organisation and to provide relevant practical advice from their respective positions."*

It is commonly accepted that effective leadership will lead to higher organisational performance and poor leadership will result in poor performance (gaard et al., 2005; Simons 2002). In its discussion of productivity drivers the Department of Labour also notes that:

> *"Leadership by example plays a strong role in creating a positive and productive workplace culture, and inspiring others to pursue those opportunities which have been identified. Leadership depth is important."*

Across all the sectors represented in the interviews, most managers are aware of their role in enhancing productivity.

> *"I am the conductor, I can put my sticks up and down and others have to perform. My performance is determined on good service, standards being met and the financials being sound".*

> *"No one likes to work for a bad manager. I have learnt from good managers I have had in the past and then applied the good things gained from them. I still watch good managers now and whenever I go into bars and restaurants I watch how their staff work. For a while I did this every month at a cafe in Auckland to learn and observe".*

Two management types emerged from the interviews: non-adaptive and adaptive. The former refers to a leadership style underpinned by a more paternalistic philosophy. The adaptive approach, by way of contrast, refers to a leadership style that emphasises things like team tasks and targets, and enhancing employees' readiness to participate in meeting overarching requirements.

It is not easy to generalise which of these approaches is superior in terms of F&B service productivity. The value of each depends on the environment in which the business is operating and the organisational culture and structure it exhibits, along with a range of other variables.

Employee Autonomy and Communication

According to many of managers interviewed, autonomy is important for hospitality employees especially when it comes to the spontaneity involved in the interaction between staff and customers. However, a culture of "controlled autonomy" also exists in some cases-especially in the larger enterprises-and is designed to control the major productivity parameters such as service standards, wastage, human resource costs. Smaller businesses, constrained by heavy and unpredictable workloads and limited resources, tend to give their staff more freedom.

Most of the managers feel that their employees have autonomy in their jobs. The difference lies in the extent to which formal methods are used to enhance communication. The larger organisations-especially hotels-have more formal consultation systems such as 'tea and scone' sessions and regular meetings where staff and management can discuss ideas and performance. As one manager from a larger organisation noted:

> *"Decision-making in an organisation this size is slower. Roll out of information is slower. No getting away from it, so the key is being smarter about how they make and implement decisions."*

Managers were asked in the web survey how they would characterise the level of input their employees have in suggesting how the workplace can be better organised? The majority of the 55 managers who responded feel that their employees have either reasonable (49%) or significant (29%) levels of input into workplace organisation; only 4% believe their employees have no or very limited input.

The ability of staff and management to communicate effectively is vital to the creation of lower-stress workplaces. Stress is associated in a number of studies with reduced work output and poor employee performance (Lo & Lamm, 2005). Although respondents in an Auckland hotel study did not report high levels of work-related stress or stress-induced illness (Lo & Lamm, 2005), other research points to the fact that unpleasant, and stressful conditions in hospitality workplaces will have a negative influence on service quality, employee productivity and business performance.

Most employees feel that their managers rarely if ever discuss productivity concepts with them. Employees of larger organisations said their relationship with their managers is often formal, with some managers being very authoritarian to the point where everything is done as dictated, and no ideas are to be expressed. Participants in the smaller organisations also came across authoritarian managers and were usually disappointed in terms of productivity outcomes:

> *"My manager does not really encourage us to give input she just wants us to do what she asks. We are not encouraged to be creative, just to follow exactly what we are told. Once I put some magazines out for customers but was told to remove them. After that I didn't use my initiative as she didn't want our ideas".*

Some employees, mostly from smaller enterprises, have managers and supervisors who are more open to staff ideas about improving productivity and the work environment:

> *"We can make suggestions to our manager and he will discuss it with people on the same shift. He is willing to try new ideas".*

Those employees in large organisations commented that being casual and/or part time, they don't have much contact with their manager, and thus *"couldn't care less as long as I get paid"*. Quality relationships between staff and management clearly have a bearing on whether the employees feel there is a good climate in which to discuss productivity, and how encouraged they feel to share ideas.

Feedback from managers on employee achievements plays a significant role for some employees in creating job satisfaction. Regular appraisals, pay rises and tips are stated as additional useful monetary indicators of productivity:

> *"One of the main drivers for me to be more productive is recognition, like one of the managers saying thank you at the end of the day."*

Smaller businesses are very positive about seeking staff input, but usually in an informal way with the owners and managers being happy to have staff input or ask staff about how things are going, often through just chatting with employees.

> *"The internal communication among the staff, from me and the staff...keeps the whole thing flowing very well and very productively...."*

The workplace climate both in terms of manager encouragement, as well as other employee attitudes, does not always encourage individuals to improve their work practices, as is illustrated:

> *"From my experience it is more like tall poppy syndrome. If you do something outside the norm everybody else may get niggled because they don't want to push themselves as much. It can be frustrating".*

Participants also admitted that making suggestions sometimes caused unwanted arguments, and organisations (especially larger) often have set ways of doing things making it hard for suggestions to be implemented. Some participants expressed doubt if their suggestions ever *"make it to the head office"*. Good communication in a team has a significant impact on the efficiency of staff and their ability to perform the *"dance"*, and helps to improve the customer experience. It is especially crucial, but hard to achieve, when there is *"not time to say please and thanks"*.

Employees in larger organisations said that communication between staff was better in smaller teams where everyone listened to each other to avoid making mistakes, which helped to reduce conflict at the workplace. These same employees explained how Christmas and social events they have experienced gave them a chance to socialise which enhanced workplace relationships and fostered greater teamwork.

On several occasions focus group participants mentioned that conflict between kitchen staff and wait staff has a significant impact on overall productivity. Wait staff often get caught in the middle when customers are unhappy with their meal and the kitchen does not appreciate the criticism. Employees think that poor English greatly affects customer service and productivity, especially in the case of phone orders. Chefs with poor English can be a particular cause of conflict among staff, exacerbating the antagonism between front of house and kitchen staff. "Customers often got annoyed and even 'grumpy' when staff speak poor English, which effectively reduces customer satisfaction".

Quality of Working Life

The majority of managers interviewed said that they feel their jobs could potentially offer a comfortable level of work-life balance, but most choose to work a large number of hours. Many spoke of their passion for what they do, rather than viewing it as work.

> *"It is full on, it is my fault though. I work 6 days, 60-70 hours a week. I do what I know, and I still really enjoy my job, especially the project work as we have so many resources".*

While several managers commented about the fact that they do not demonstrate good work-life balance themselves, they stated that they are highly aware of the importance of it for their staff productivity.

> *"Work-life balance is very important-they [staff] are rostered off with a weekend off once a month, and are not allowed to come in. We try to roster the days they want off".*

The managers who responded to the web survey were asked if they think hospitality is an attractive career. Of the 53 participants who responded, 75% stated that they did feel it is attractive, with only 25% disagreeing.

Employees were less sure about the ability to manage work-life balance in a F&B services role, and less certain about F&B offering attractive career options. Working too many shifts can be a "novelty" at first, but inevitably results in one being overworked and stressed out. Most employees felt it is harder for permanent employees to achieve work-life balance, making some employees frustrated and leading them to seek less demanding, more motivating, and better paid work.

Employees from a range of organisation types said that hospitality work is *"hard and tiring and poorly paid"*. Negative comments about hospitality work also referred to being paid only minimum wage, working long hours, not getting paid overtime, rarely getting tips and bonuses, stressful work environment, and having to deal with negative customers. Uncertainty over work-flows was often commented on:

"If we are short of staff everyone has to do anything. I have done room service, a la Carte, as well as cashier. When we are not busy it is a totally different pace".

The overwhelming majority of employees say they will not be working in hospitality in the future. Those studying at university plan to quit their hospitality jobs after completing their studies. Among the reasons for quitting are: it's not a *"real"* career; it's not challenging enough; it has negative customers, and low pay. Employee dissatisfaction with high levels of formality at the workplace, as well as their desire to change their job is also apparent in larger enterprises.

"Never took it to be a career, just the hours suited".

There were a number of participants from a range of (mostly small) organisations who genuinely do enjoy working in the industry. These individuals are more likely to be working in establishments where their contribution is sought and valued, and display the attitude and enthusiasm for the industry demonstrated by the managers interviewed. For some employees flexibility in their work environment was also a real positive:

"I have flexibility at work; they are not too strict as long as you do your job properly. The cafe is busy but there are times when there is not too much pressure".

Some employees in larger organisations also believe that hospitality is becoming more professional, offers more opportunities, and people are *"starting to get paid more".*

Support Networks

As the Department of Labour (2004) notes:

> *"Firms do not operate in isolation and there are significant productivity gains to be achieved by improving the exchange of knowledge, information and ideas through both formal and informal networks".*

Those employers that completed the web survey use a variety of sources to help build managerial capability. The web survey participants were asked where they went to for help as an employer/ manager. The two most popular sources are friends and personal networks (73%), followed by accountant or lawyer (45%). The Department of Labour ranks third as a source for those seeking assistance (43%).

Productivity is a topic for discussion for some managers with colleagues in their internal and external networks. External parties were often identified as trustworthy sources with which to discuss productivity, particularly regarding new products and process innovations.

About a third of the managers said that they do not really discuss productivity, especially beyond the walls of their own establishment.

> *"Productivity is not talked about in NZ like it is overseas-more a cost balancing act here. The industry is not so good at discussing productivity."*

For employees networks also play an important role. In many cases ethnic or friendship networks are important in the hiring process-with recommendations for existing staff playing a role in the attraction of new workers. Within firms the 'social capital' associated with close knit work teams was also often mentioned-with effective team work seen a great way to learn new skills and create a more enjoyable work environment.

Hiring & Training

According to Fair and Brook (2001), businesses should "never underestimate the contribution to productivity that ongoing everyday training makes". At the same time the Department of Labour (2004) stresses.

> *"Skills shortages choke off growth potential. The knowledge, ability and skills of workers contribute to workplace productivity. Ongoing investment in foundation, technical, supervisory and managerial skills, together with improvements in work organisation, can help improve the productivity and performance of New Zealand firms."*

The link between training and productivity is reflected in recent comments by Bruce Robertson, CEO of the Hospitality Association of New Zealand.

> *"One of the other keys to improved productivity is training. Three trained staff will always outperform four untrained staff and that applies right throughout the industry".*

Research has shown that effective hiring and training is linked to a better ability to meet the needs of a target market (Shaw & Patterson, 1995), greater organisation commitment (Roehl & Swerdlow, 1999), more effective use of new technology, and enhanced work outcomes.

A firm can only achieve higher productivity by employing the right individuals with the capacity and desire to contribute to its goals.

> *"Good hiring creates a team; good communication creates a feeling of involvement, that it's more than just a job..."*

Managers were asked in the web survey to identify the key characteristics or skills that they look for when recruiting staff. The two most commonly mentioned characteristics are honesty/loyalty, and experience (30% each). Personality and ability are sought by 23% and 20% of managers respectively.

Increasingly, these managers are recruiting from international sources and new migrant groups-this can lead to language difficulties. Managers were asked in the web survey about the adequacy of numeracy, literacy and language skills among applicants.

Over 80% of managers feel that overall, applicants do have adequate language and literacy skills, although only 66% test language and 59% test literacy skills when hiring. Only a third of managers test for numeracy skills when hiring.

Training can positively impact on employee productivity, promoting customer satisfaction and sustaining competitive advantage. Training is associated with improved job satisfaction and intention to stay, and with reduced workplace problems and staff turnover (Poulston, 2006).

Most managers interviewed for this study believe training has very positive and direct impacts on productivity improvement. Some rank it the number one factor and most of them describe it as having *'important'*, *'crucial'*, *'huge'* and *'massive'* importance. On-the-job, in-house, and self-training are the most common forms of training in the industry. Constrained by cost, many the interviewees from smaller enterprises expressed their inability and reluctance to employ outside trainers, although a couple of them talked about using outside trainers in specialist areas.

These findings are confirmed through the web survey which indicates that 88% of managers say the most common form of training engaged in is informal and on-the-job. This is followed by formal in-house training (40%) and training by an external provider (31%). Over half the managers think the training they are providing to their staff matches their needs. However, constrained by time, money and business pressure, a number of managers expressed their concern about the lack of training they were able to provide.

Tertiary/private training is considered to be too theoretical and as not meeting the industry's requirements well. Most of the managers interviewed, and 60% of web survey participants, feel formal tertiary/private hospitality qualifications do not adequately prepare people to enter employment in their workplace.

> *"I think on the job training is very important-it is hard to take a student from tech and put them in this environment".*

A theme of *"wanting to train them our way"* emerged amongst some of the managers interviewed. Some managers expressed annoyance at needing to at times 'untrain' those that had been trained to do jobs one way. Employees had generally undergone some type of training when they first started their job; and the majority had on-the-job training provided by other staff and/or the manager. During on-the-job training, participants appreciate the opportunity to see senior staff *"showing tricks of the trade"* in the context of the workplace, and to ask questions. Employees said training provided by the manager rather than senior colleagues was most valuable, as other staff often *"forgot little things"* or deliberately *"withheld information"* on the best ways to do things. Some didn't have either formal or on-the-job training and were forced to follow a *"sink or swim"* approach-*"You are given a till and a bag of money and told to go for it".*

Employees in larger organisations tended to have taken formal induction courses. During induction training, participants watch videos, are given manuals and introduced to key procedures. Research shows that employees of small and medium enterprises are less likely to have the opportunity to take part in this type of training than those working in large organisations Milne et al. (2004).

Employees in the focus groups generally believed that *"training helps to access jobs",* but were less certain about its ability to provide access directly to managerial level jobs. Those in small organisation focus groups felt that there is a *"gap between education, hospitality and employers' expectations"* and most expressed some scepticism about the ability of tertiary providers to really prepare people for the F&B workforce.

Turnover & Retention

As one F&B manager Interviewed Noted: "Employee retention is important for consistency of products and customer service standards, so customers know what to expect".

Hinkin and Tracey (2000) list the following productivity-related costs of staff turnover: pre-departure productivity loss; learning curve; errors and waste; supervisory and peer disruption. Turnover can also negatively affect productivity within interdependent worker groupings and teams. However not all staff turnover has negative implications.

Moderate turnover can help to replace unproductive employees, enable career progression, enhance job and employee matching and is therefore viewed as enhancing productivity by some (Ilmakunnas et al., 2005). A recent study of Auckland hospitality workers (Poulston, 2006) indicated short lengths of employment are common (87% have been in their current job fewer than five years), and that the majority work part-time. The latest survey conducted by the Restaurant Association (March 2006) indicates an average annual staff turnover of 29%, up from 26% in 2005.

A New Zealand study of productivity found that high-involvement work practices and employee retention were strongly associated with strong enterprise performance (Guthrie, 2001). While turnover increases exposure to disruptions associated with the loss of employees in high involvement work practices this damage is greater than in settings where processes are more automated, for example the preparation areas of fast food establishments.

Several managers stated in the interviews that they have real problems with staff turnover, tending to attribute much of the problem to the tight labour market rather than to factors that are directly within their control. Of the managers who completed the web survey question asking whether staff turnover affects productivity in their business, 54% feel it does, 29% feel it does not, and 17% responded 'not applicable'. The need for training new staff was seen as having the greatest negative effect on productivity. The time and cost involved are also detrimental effects stemming from staff turnover.

> *"Turnover is costly and involves extra training-there's also a trust issue with long term staff".*

Large hotels are the most proactive in developing strategies to cope with turnover and to increase retention. Such strategies include succession planning, valuing internal communication to develop teams, and offering incentives for staff to stay. Training and flexible hours are also seen to be important in improving retention.

Overall, employees agreed that new staff often reduce productivity, because they are inefficient in their job, and because it is a burden on experienced staff to train someone new. Some employees, however, give new staff credit saying they usually are the most willing to learn.

Longer term staff not only have more industry experience, which helps to increase overall productivity, but also a stronger relationship with the kitchen staff, seen by many as prerequisite for having a *"good night"*.

Technology

In its discussion of productivity drivers the Department of Labour has identified that:

> *"innovation is a key part of raising workplace productivity. The appropriate introduction of advanced technology is linked with higher productivity, greater market share and employment growth".*

The introduction of advanced technology has been cited by several authors as an important and positive influence on productivity. Positive impacts however do not simply 'appear', rather they stem from the competent use and exploitation of IT capabilities (Sigala, 2003).

Technology has been found to 'support employees, enhance the quality of service, improve efficiencies, gain competitive advantage, maintain relationships with customers and increase profitability', while poor investment in technology (along with poor training) is likely to reduce profitability.

Several studies have shown that SMEs in the hospitality sector have special difficulties in maximising the productivity potential of information technologies. Such enterprises are more likely to recognise the limitations of technology and emphasise the importance of 'the human touch'. At the same time SME owners and managers are more likely to have to out-source aspects of IT maintenance and may feel more removed from its application to the business setting.

The web survey found that 89% of managers have introduced new technology in the last two years with the view to improving productivity in their workplace.

The interviews revealed a similar trend. A range of technologies have been introduced by managers, primarily in the front of house area with computers being the most common purchase. Smaller businesses have a slower uptake of 'cutting edge' ICT devices such as wireless personal digital assistants and text message ordering.

Both managers and employees are divided on the issue of technology and productivity. Some think it enhances staff efficiency, saves time and improves productivity. Others say it in fact creates inefficiencies when it breaks down or requires updating.

While technology was discussed as an important contributor to productivity, most of the technology identified as bringing real efficiency gains is in the front of house environment. In particular, technological innovations are central to streamlining ordering and point of sale processes.

Chapter 3

Planning and Management of Food Services

In the past few decades, people ate in restaurants occassionally to celebrate a special event such as an anniversary, a birthday, or an achievement. It was an outing to look forward to, and if one indulged, it did not matter as these outings were rare. Today, the scenario is different. Eating out has become a way of life. Education and employment has taken many of us away from home, and the mother's role now has an added responsibility of contributing to the family income. Modern day compulsions have made eating out a necessity. No longer does one find time for the traditional fare of yesteryears and depends on the caterer for the following:

1. Food for festivals and celebrations
2. Meals at the work place.
3. Ready-to-eat meals picked up on the way home from work
4. Snacks and sweetmeats for daily consumption
5. Preserves, pickles, papads, etc.
6. All meals served in institutions such as hospitals, school/college cafeteria, mess or dining hall, and boarding schools.

The number of reported cases of diabetes, hypertension, obesity, heart attacks, etc. is on the rise and so is the number of meals consumed away from home. This is not surprising because if one indulges practically everyday, it is bound to result in ill-health because of malnutrition. The caterer's role has become more significant as the responsibility now lies with the caterer for planning nutritionally adequate meals. Menu planning is the key to overcoming this problem.

Definition: Menu planning is defined as a simple process which involves application of the knowledge of food, nutrients, food habits, and likes and dislikes to plan wholesome and attractive meals.

The caterer who is responsible for providing meals has to decide on various aspects, such as:

1. Menu
2. Serving size
3. Food cost
4. Suppliers and quantities to be purchased
5. Standardized recipes to be followed
6. Type of service
7. Meal timings
8. Clientele.

The aim of menu planning is to:

1. Meet the nutritional needs of the individuals who will be consuming the food
2. Plan meals within the food cost
3. Simplify purchase, preparation, and storage of meals
4. Provide attractive, appetizing meals with no monotony
5. Save time and money
6. Minimize overhead expenditure, i.e., fuel, electricity, water, labour.

Menu planning is the most important aspect of planning and organization in the food industry. It is an advance plan of a dietary pattern over a given period of time.

Menus are of the following types.

Banquet menus: These are special menus for banquets or functions.

Institutional menus: Hospital menus, boarding school menus, and industrial can-teen menus.

Menus may be cyclic which means they are compiled to cover a specified period of time.

The length of the cycle may vary and is decidedupon by the management. A number of menus are set and repeated. They are often modified to take into account variations which may arise for a number of reasons.

Factors Influencing Meal Planning

Many factors influence the acceptability of a meal. Customers select what appeals most to them from a menu card based on individual likes and dislikes, budget, popularity of items, etc. However, while planning meals the following factors need to be considered.

Nutritional Adequacy

The most important consideration in menu planning is to ensure that the meal fulfils the nutrient needs of the individual consuming the meal. For example, if the meal is planned for an industrial worker, it must meet the RDAs for that age group. Foods from all basic food groups should be included in each meal so that the meal is balanced and nutritionally adequate. Nutrient needs may be modified for hospital diets (therapeutic diets).

Economic Considerations

The spending power of the clientele has to be kept in mind and meals have to be planned within the budget. Low cost nutritious substitutes should be included in the menu to keep the costs low. The food cost should be maintained, if the organization has to run profitably.

Type of Food Service

Menus should be planned in relation to the type of food service, whether it is cafeteria, seated service, buffet, etc.

Equipment and Work Space

The menu should be planned keeping the available equipment and work space in mind. Deep freezers, refrigerators, grinders, dough kneaders, deep fat fryers, boilers, etc. should be adequate.

Leftover Food

An effective manager should consider as to how leftovers could be rotated to obtain maximum profit. Adequate storage space and hygienic standards should be ensured to minimize the risk of contamination and spoilage of food.

Food Habits

Food habits of the customer is another important criteria which needs to be considered as food served has to be acceptable to the customer. Special attention should be paid when a particular type of community is catered to. Religious considerations should be known to the meal planner.

Availability

Some fruits and vegetables are seasonal. During the season the cost is reasonable and quality is better. Today, practically all fruits and vegetables are available throughout the year because of advanced preservation technology. However, seasonal fruits and vegetables should be given preference. Regional availability influences menu planning. For example, fish and sea food is fresh and cheaper in coastal areas.

Meat Frequency and Pattern

The meal timings and number of meals consumed in a day, whether meals are packed or served at the table, also influences the selection of food items on the menu. The age, activity level, physiological state, work schedule, and economic factors need to be known before planning meals for institutional catering.

Variety

This is one of the most important considerations while planning meals. A variety of foods from the different food groups should be included. The term variety means

1. Variety in food ingredients
2. Variety in recipe
3. Method of cooking
4. Colour, texture, and flavour
5. Variety in presentation and garnish.

A meal should look attractive and be appetizing. A judicious blend of flavours, attractive colour combinations, and different textures make food enjoyable and interesting. The method of cooking used for different items on the menu should vary.

All animal proteins Pulses, nuts, and oilseeds.

Protein, vitamin and mineral rich Protein, vitamin, mineral, fibre, oils.

For example, two deep fried items would make the meal heavy. Simple processes such as fermentation and sprouting not only contribute to improved flavour and digestibility, but also enhance the nutritive value of the meal.

A well planned meal which is nutritionally adequate would have a good satiety value and prevent the occurence of hunger-pangs before it is time for the next meal. The nutritional adequacy of a meal in an a la carte service depends on the food choices made by the customer. It is the duty of the caterer to offer adequate, nutrient dense foods to the clients, to choose from.

Planning Balanced Meals

Meal planning involves proper selection of food to ensure balanced meals. We have also read that food can be classified on the basis of its source, the nutrients present in it, or on the basis of its functions into 3-11 food groups. These food groups help us in planning balanced meals which supply all essential nutrients. In this chapter we will study the three basic food groups classified on the basis of functions performed by nutrients as this is the simplest way to ensure adequate nourishment to the body.

The three main functions performed by food are:

1. Providing energy
2. Body building and maintenance
3. Regulation of body processes and protection against infection.

On the basis of functions performed, food is classified into the following three groups.

1. Protective/regulatory foods
2. Body building foods
3. Energy giving foods.

Steps In Planning Balanced Meals

1. Collect information regarding the customer with respect to
   - Age
   - Gender
   - Activity level
   - Religion
   - Socioeconomic background
   - Food habits
2. Check the RDAs for energy and proteins
3. Prepare a food plan, i.e., list number of servings from each food group to meet the RDA
4. Decide on number of meals
5. Distribute servings for each meal
6. Select foods within each group and state their amount

7. Plan a menu
8. Cross check to ensure that all food groups are included in requisite amounts.

Using the above steps, plan a balanced diet for a day.

Example: Planning balanced meals for college students residing in a hostel.

Calculating the Nutritive Value of a Recipe

Using the food composition table, follow the steps given here to calculate nutritive value of a recipe.

1. List the ingredients used and their quantities in the recipe.
2. Prepare a table with the following blank columns, and fill up the ingredient and quantity column from the recipe.
3. Refer to the food composition tables for the nutrients present in 100 g of edible portion of each ingredient.
4. Calculate the nutrients present for the quantities used in the recipe.
5. Weigh the finished product to know the total yeild.
6. Divide these values by the number of portions to know the nutritive value per portion.
7. For general calculations, do not include salt, spices, baking powder, stock, or ingredients which are used in very small quantities (less than 10 g) except for sugar and fat. Baking powder and salt are calculated for their sodium content, and not for their proximate principles, for sodium-restricted diets only.

Special Nutritional Requirements

Pregnancy

Every infant should have the right to begin life with a full-term healthy body and receive the advantage of mother's milk. For this to happen, every would-be mother should take special care of her nutritional needs. Pregnancy is a period of remark-able anabolic activity. A 3.2 kg infant develops in 9 months of pregnancy from the nourishment received from the mother. If the diet is deficient during these months, the foetus draws upon the maternal reserves. A poor diet will ultimately affect both the infant and the mother, and may lead to complications of pregnancy such as pre-mature birth and low birth weight.

Weight gain: A healthy woman is expected to gain 16-24 lbs in the 3 trimesters of pregnancy of which the full-term infant weighs 7 lbs. The remaining weight is due to increase in tissues, an expanding blood volume, and energy stored in the form of fat as a calorie reserve.

Energy: The energy needs increase during the second and third trimesters of pregnancy. An additional 300 kcalories take care of the increased demands of pregnancy. Calorific value of food is adjusted according to the weight gained.

Protein: The protein requirement increases by 15 g for the synthesis of foetal and maternal tissues.

Minerals: The calcium requirement is 1,000 mg and iron is 38 mg for the formation of foetal bones and teeth, and formation of blood respectively. The absorption of minerals improves because of the increased requirement. Iodine in the form of iodized salt, helps in protecting the mother and child against goitre and cretinism.

Vitamins: With an increase in calories, the need for vitamins B_1, B_2, and niacin increases as they are needed for release of energy from carbohydrates, fats, and proteins. The need for folic acid and vitamin D also increases.

Nutrient dense foods should be selected to meet the extra demand for proteins, calcium, iron, and vitamin D and B-complex vitamins.

Lactation

Mother's milk is the most nutritious food designed for an infant. The nutritional needs during lactation are greater than the needs during pregnancy as the mother's body has to supply all nourishment to the rapidly growing infant. The conversion of dietary protein to milk protein is only 50% which means 2 g of good quality pro-tein is converted to 1 g of milk protein.

Energy: Calorie requirements increase by 550 kcal during the first six months of lactation, followed by a marginal decrease during the next six months with 400 kcal being adequate to meet the additional demands.

Proteins: An additional intake of 25 g in the first six months and 18 g in the next six months is adequate.

Minerals: Calcium requirement is 1,000 mg which is necessary for synthesis of milk. Additional iron is not prescribed as milk is a poor source of iron.

Fluid: The intake of fluid increases during lactation.

The diet should be nutritious, easy to digest with restrictions on strongly flavoured vegetables and spicy food.

Human milk is the natural food for the infant. It is safe and convenient as it does not involve sterilizing bottles and preparing formulas throughout the day. At the same time, it is easy to assimilate, has the correct temperature, and gives a safe and secure feeling to the infant, and sense of satisfaction to the mother.

During the first few days after delivery, colostrum is secreted which is not mature milk but a substance richer in protein and vitamin A. Colostrum is secreted in small quantities but is valuable as it increases the resistance to certain infections during the first few months of life.

The intervals of feeding should be fairly flexible instead of following a rigid schedule or a self-demand schedule. For premature and low birth weight (weight less than 2.5 kg) babies a fixed schedule is preferable. Milk is deficient in iron, vitamin C, and vitamin D. The baby is born with stores of these nutrients which suffice for 3 months. From the third month onwards, supplements should be gradually added to provide those nutrients which are not sup-plied by milk.

From the fifth to ninth month onwards, the infant should be weaned by substituting a cup feeding for a breastfeeding. This change should be gradual and at intervals till the infant is weaned from the breast to the cup before the age of one year.

Nutritional requirements: The next best substitute is cow's milk or modified buffalo milk. Buffalo milk is modified as it con-tains a larger percentage of protein, fat, calcium, and energy as compared to human milk and a lesser percentage of sugar. It is modified by partially skimming it, diluting it with water, and adding little sugar so that its composition resembles human milk.

Supplementary foods during the first year: The age at which supplementary foods should be included depends on a number of factors such as literacy of mother, economic factors, and time available. Supplements if not prepared, stored, and fed to the infant in hygienic conditions is one of the main causes for diarrhoea and gastrointestinal upsets. While introducing a new food the following points should be borne in mind:

1. Introduce one food at a time till the infant's system is used to the new food. Season with salt
2. Give a teaspoonful in the beginning

3. Never force-feed an infant
4. Start with a thin smooth consistency initially
5. Once the infant is used to supplements, introduce variety
6. Follow hygienic practices and do not give any leftover food or drink to an infant.

Choice of food supplements for infants:

1. Egg yolk
2. Strained cereals
3. Strained soup
4. Soft khichdi
5. Rice and dal
6. Pureed vegetables
7. Ready to eat infant cereals
8. Stewed fruit
9. Fruit juice.

Childhood

During these stages rapid growth takes place. There is an increase in height and weight because of increase in the bone and muscle mass.

Children use up a lot of energy in playing. They normally carry their lunch, which should be nutritious and interesting, or monotonous, in which case the packed lunch box comes back home unfinished or unopened. Special efforts should be made so that they do not meet their energy needs from junk food.

More energy per kg body weight is required for:

1. Rapid growth which takes place in this age group
2. Enhanced physical activity
3. High BMR as compared to an adult.

Proteins: Good quality proteins should be included to take care of body building and maintenance of tissues. Milk proteins are complete proteins and provide calcium as well.

Carbohydrates and fats: They provide calories, and spare proteins from being oxidized for energy. Refined carbohydrates and poor dental hygiene are the reasons for dental caries in children. The consumption of fruits and vegetables, both cooked and uncooked, should be encouraged. Junk food and aerated beverages should be discouraged.

Vitamins and minerals: If the diet is well planned keeping the principles of menu planning in mind, supplements may not be necessary.

Fluid:Adequate fluid is necessary, specially in active children who play outdoor games and sweat a lot. Fluids should not kill the appetite, which is seen when children are thirsty and drink a lot of liquid just before a meal.

Points to be considered:

1. Regular meal timings
2. No nibbling low nutrient dense snacks between meals
3. Do not force feed, let the normal appetite return
4. Meal timing should be pleasant
5. Food should be appetizing and attractively served. Finger foods should be preferred for children
6. Packed meals should not be messy to eat
7. Attractive colours and shapes appeal to children, and with a little imagination eating green vegetables and drinking milk could become more interesting.

Adolescence

Adolescence is a period of physiological stress for the body because of extremely rapid rate of growth. The appearance of sex characteristic is also accompanied by mental and emotional changes. The diet plays a crucial role in promoting growth, hence the RDAs for all nutrients are high during 13-18 years of age. Nutrients of particular importance are carbohydrates and fats for energy and proteins, iron and calcium for body building. During adolescence the basal metabolic rate (BMR) accelerates once again because of the growth spurt and other hormonal changes which take place. The BMR of boys is higher than that of girls because of more muscle tissue. Food habits change drastically because of peer pressure, maintaining one's figure and weight, skin problems, and the newly found independence.

Energy: Energy needs are high because of higher physical activity and BMR. Sports and aerobic exercises increase the energy requirements for sports people.

Proteins: A generous intake of high quality proteins is necessary for increase in muscles mass and skeletal development. A deficiency of calories and proteins can affect one's optimum height increase and resistance to infections.

Girls grow rapidly between 11-14 years and boys grow rapidly between 13-16 years. Initially, girls are taller than boys between 11-12 years. But later the growth rate is more in boys. Weight gain is a common problem in this age group which is tackled differently by both the sexes. Boys prefer exercise as a means of losing weight and body building while girls prefer going on a diet.

A deficiency of calcium/vitamin D during infancy or childhood results in rickets (deforming of bones). The weak bones cannot withstand the weight of the body and bend causing bow legs or knock knees. Once malformed, bones cannot be straightened. The effect of rickets is seen in adolescents and adults. Along with weight, emotional stress has an adverse effect on health and diet. Friction between parents and adolescents is another area of tension.

Snacks are the all-time favourite food, which often forms a large percentage of energy intake, favourite snacks being burgers, pizzas, pastries, ice creams, french fries, pav wada, pav bhaji, popcorn, and South Indian snacks. Milk is substituted by tea/coffee (which is more of a social need) or aerated beverages. This trend indicates a high consumption of refined flour, hydrogenated fats, and potato. South Indian snacks are a better option as they are made from a cereal-pulse combination. In general, the diet is deficient in the protein food group and protective food group. Food such as milk, green leafy vegetables, yellow fruits and vegetables, citrus fruits, and whole grain cereals are deficient in the diet of adolescents. Deficiencies during infancy and childhood may leave their scar in adolescence.

Old Age

Anorexia nervosa: This is a disorder seen in adolescent girls concerning figure and weight control. The onset of puberty results in widening of the pelvis and deposition of subcutaneous fat creating a psychological problem in figure conscious young girls. They resort to crash diets for weight loss. This self-induced starvation to attain a slim figure leads to weight loss, loss of muscle tissue, a low BMR, and serious health problems.

Bulimia: It is another disorder in which the appetite increases drastically followed by induced vomiting to throw out whatever has been consumed. Bulimia creates both emotional and physical problems and is also known as the gorging-purging syndrome. Both disorders are mainly seen in adolescent girls from affluent families.

Nutrition in old age or geriatric nutrition is gaining importance as longevity has increased and 65 plus is the fastest growing segment of our population. Aging is an individual phenomenon depending to a large extent on the health and nutrition status of an individual throughout his life span. People grow old at different rates and in different ways. Aging continues from birth throughout life.

Apart from the biological changes which occur with age, psychological and social changes are also seen. The body functions slow down, there is increasing social stress, and isolation. The elderly are often lonely, restless, and unhappy. They feel they are a burden and cannot contribute physically or monetarily to the family.

The following changes are seen in old age.

Body composition: As age advance, adults lose muscle mass and gain fat. Since active muscle is lost, BMR reduces with age. Bone loss occurs leading to a reduction in height. Collagen content increases giving rise to loss of elasticity in blood vessels, joints, and skin, seen as easy bruises, painful joints, and a wrinkled skin.

Sensory changes: The five senses of sight, hearing, taste, smell, and touch become dim with age. Old people cannot see without bright light, hearing is impaired affecting their social behaviour, and the other senses are less acute.

Digestive functions: Loss of teeth and ill-fitting dentures make it necessary to serve soft well-cooked food. Decreased secretion of saliva makes food difficult to swallow. Heartburn, diarrhoea, or constipation are other complaints.

Metabolic changes: Changes in metabolism, specially glucose tolerance and anaemias, are sometimes seen.

Osteopenia and osteoporosis are seen resulting in excessive bone loss, and fractures due to normal stress is another common finding. Because of age-related disabilities, the elderly should not live alone as they require assistance for all activities and company to overcome psychosocial changes in their life.

Nutritional requirements: The main features of the diet are as follows:

1. A low calorie diet of 1,000-1,500 kCal, as there is a reduction in the basal metabolic rate (BMR).
2. Good quality protein, calcium, and iron. Vitamin A, vitamin C, and folic acid to prevent anaemia.
3. Food should be easy to chew, liquid preparations should be included in each meal as they are light and easy to digest.
4. Hot meals should be served attractively as appetite is low.

5. Five small meals should be preferred to a three meal pattern.
6. Gas forming foods and strongly flavoured vegetables such as bengal gram, cauliflower, cabbage, and onion should be avoided.
7. Fibre rich foods such as whole grains, green leafy vegetables, pulses, and fruits are nutritious and provide fibre necessary for normal elimination. They should be incorporated into the diet gradually to prevent gastric irritation and gas.
8. Fluids: Ample fluids are necessary to flush out toxic wastes. The kidneys function at a slower pace with age, but fluids help in excretion of toxins.
9. Low calorie density and high satiety value foods should be included.

Effect of Quantity Cooking and Processing on Nutrients

Almost all foods consumed today need some form of cooking and processing before it is fit for service and consumption. Fruits and vegetables used in salads or for chutney are consumed uncooked. The nutrients we receive from the meals we consume depend to a large extent on cooking and processing practices which are being used. While some amount of nutrient loss is inevitable, cooking has many benefits which are listed below.

Benefits of Cooking Food

1. Cooking increases palatability
2. Cooking makes food easier to digest by destroying anti-digestive factors such as trypsin inhibitor in soya beans
3. Pathogenic microorganisms are destroyed
4. Shelf life is increased by destruction of spoilage organisms and denaturation of enzymes
5. The appearance of food improves, e.g., cooked meat versus raw meat.

Common Food Processing Techniques

Food prepared in large quantity in Institutional Kitchens or in food processing plants is more prone to loss of nutrients, if adequate care is not taken to retain or preserve the nutrient.

This is because if food is cooked in bulk, the pre-preparation begins hours in advance, for example, vegetables have to be cut in advance and if these are not blanched and refrigerated to inactivate enzymes, oxidative losses of labile vitamins will continue at room temperature. Apart from nutritive value, the crisp texture of salads is also lost and phenol containing vegetables will discolour and turn brown, making the dish unattractive and unappetizing.

Effect of heat on nutrients: Cooking has beneficial effects on carbohydrates because of gelatinization of starch, favourable browning reactions such as Maillard reaction and caramelization of sugar which gives colour and flavour to food. Proteins too take part in Maillard reaction along with sugar. Enzymes which catalyse undesirable enzymatic reactions in fruits such as apple and pears, and vegetables such as potato and brinjal are inactivated on blanching or cooking these foods. Enzymes which hasten oxidative destruction of vitamin C or ascorbic acid are denatured by blanching. Proteins get denatured by heat. The chemical reactions that take place when oil is heated continuously during deep fat frying bring about hydrolysis, oxidation, and polymerization of the oil.

Triglyceride

The release of moisture, high frying temperatures of 160°-190°C, presence of carbonized crumbs in the oil, and oxygen from the atmosphere during frying brings about oxidation of the oil. Repeated use of the frying medium forms thermal and oxidative products which can cause gastrointestinal irritation and destruction of vitamins. These products undergo polymerization and increase the viscosity of the oil. The oil darkens in colour, has a lower smoke point, and foams when used for frying. Such oil should be discarded. Fat soluble vitamins dissolve in fat used for deep frying specially if food to be fried is not well coated.

Effect of alkali: Alkali is used during cooking and processing to soften vegetables, make pectin soluble, and dissolve hemicellulose. It is also used as lye,sodium hydroxide) to peel vegetables during processing. A pinch of sodium bicarbonate added to green vegetables helps in brightening the green colour. However, B-complex vitamins and ascorbic acid are destroyed in an alkaline medium. The use of alkali to hasten the cooking process for vegetables and pulses should be discouraged. Excessive cooking in an alkaline medium not only destroys vitamins, but makes the texture mushy and gives a soapy taste to the product.

Effect of acid: An acidic medium while cooking helps preserve water soluble vitamins and retards enzymatic browning of certain fruits and vegetables. Vegetables and pulses take a longer time to cook in an acidic medium as acids precipitate pectin and hardens vegetables.

Effect of washing and soaking: While preparing food, water soluble vitamins and minerals leach out into the cooking or washing water. These losses can be minimized by washing the uncut fruit or vegetables and not soaking the cut vegetable in water. Soaking grains or pulses is beneficial as soaking increases digestibility and reduces cooking time.

Effect of sprouting and fermentation: Soaking whole grains overnight in water and tying them in a muslin cloth to allow them to germinate has many beneficial effects.

1. In sprouted grains, the dormant seed becomes active and synthesizes vitamin C.
2. Partial breakdown of carbohydrates, proteins, and fats begin, making it easier to digest.
3. The bioavailability of nutrients especially calcium and iron increases.
4. The active seeds synthesizes vitamin C or ascorbic acid and thiamine, riboflavin, and niacin content increases.

Exposure to air or oxidation: Exposure of finely divided foods to oxygen of the air reduces the vitamin C content by oxidation. The enzyme ascorbic acid oxidase is released when fruits and vegetables are cut. The enzyme activity is temperature dependant and can be inactivated by blanching, or by storing cut fruits and vegetables at refrigeration temperatures or by adding acid.

Vitamin A is destroyed on exposure to air. The colour of cut carrots (carotene) fades due to oxidation and B-complex vitamins are also affected.

Milling: Whole meal flour contains all nutrients present in the grain. In flour with 100% extraction no nutrients are lost. Low extraction flours (45% extraction) are light in colour and are mainly starch with some protein and fat. Approximately 70°/0 of all B-complex vitamins, minerals, and dietary fibre present in the whole grain are lost during milling. Polished rice (the form in which rice is consumed) loses 75% vitamin B_1 or thiamine, while parboiling helps in retaining some of the vitamin.

Cooking and processing practices vary widely from one region to another, hence no authentic information on exact losses can be known. While cooking has both adverse and beneficial effects, proper practices can minimize the adverse effects and maximize the benefits so that food can become more whole-some and safe.

Chapter 4

Marketing of Food Services

Menu Driven Business Planning

So you want to open a restaurant? Many people come to me with their plans, and ideas after they have decided to build or open a restaurant. It is my responsibility to assist them in their planning process and determine how clearly they have thought their concept through. I ask them to show me a menu, and if their initial response is 'I haven't getting that far in my planning process yet,' my response is "You haven't started the planning process yet". A menu is the foundation of any restaurant; Guests will support or avoid a restaurant for its food. Starting with a preliminary menu is a simple and basic approach to restaurant development. Begin with a menu, and you are light years ahead in the restaurant development process. A menu will tell you and your Guest what you are trying to be as a business, and greatly enhance your chances for success.

I view a menu for content, image and pricing. Content (the actual items on the menu) will dictate service staffing needs, level of culinary experience and type of management required. Who will be doing the cooking, do they have experience in this type of food, and how much are you paying them? Image is how the Guest will perceive the menu. Menu image helps define the targeted clientele and which other restaurants this operation would be competing with. Are the content and image of the menu appealing to your desired clientele? Pricing helps determine a potential restaurant's competitive placement. Is the pricing for the type of food offered competitive with other's in the market area, and does it permit the ability to manage a profitable food cost? Pricing sets the Guest's expectations in terms of food and service quality. This perception will, in turn, help define appropriate staffing levels. The budgeting process can now begin. Analysis of menu content, image and pricing will tell prospective restaurant operators whether their concept is appropriate for a certain market area.

With a preliminary menu in hand, a prospective operator can target a location that will be convenient and appropriate for their desired clientele. Once a site or facility is selected, sales volumes can be projected based on number of seats, menu pricing and the competitive business analysis. With projected sales volumes, how much an operator can spend to acquire, remodel or build a facility is determined. Leases and/or purchase agreements can now be negotiated. With a clear menu, competitive analysis, sales forecast and development budget, financing can realistically be sought. A business plan can be derived which, if taken to potential investors will demonstrate what type of return they can anticipate on their investment.

Any restaurant business plan must begin with a menu. A proposed menu provides the basics for many questions that must be answered during the restaurant development process. It creates an image of the restaurant, identifies targeted clientele, and defines the proposed restaurant's competition. A preliminary menu allows a sound basis for business budgeting, tests potential profitability, and dictates the development dollars required for a facility. Most importantly, beginning the business planning process with a menu maintains the focus of ownership on the importance of food and the impact it has on the success of the restaurant.

Pasta Restaurant Marketing Plan

Sigmund's Gourmet Pasta will be the leading gourmet pasta restaurant in Eugene, or with a rapidly developing consumer brand and growing customer base. The signature line of innovative, premium, pasta dishes include pesto with smoked salmon, pancetta and peas linguini in an alfredo sauce, and fresh mussels and clams in a marinara sauce. Sigmund's Gourmet Pasta also serves distinct salads, desserts, and beverages. Sigmund's Gourmet Pasta will reinvent the pasta experience for individuals, families, and take out customers with discretionary income by selling high quality, innovative products at a reasonable price, designing tasteful, convenient locations, and providing industry-benchmark customer service. In order to grow at a rate consistent with our objectives, Sigmund's is offering an additional $500,000 in equity.

Existing members will be given the first option to subscribe to the additional equity to allow each of them to maintain their percentage of ownership. The portion not subscribed by existing members will be available for prospective new investors.

Situation Analysis

Sigmund's Gourmet Pasta is close to entering their second year of operation. The restaurant has been well received, and marketing is now critical to its continued success and future profitability. The store offers an extensive offering of gourmet pastas. The basic market need is to offer individuals, families, and take out customers fresh, creative, attractive, pasta dishes, salads and desserts. Sigmund's uses homemade pasta, fresh vegetables, and premium meats and cheeses.

Market Summary

Sigmund's Gourmet Pasta possess good information about the market and knows a great deal about the common attributes of our most prized and loyal customers. Sigmund's Gourmet Pasta will leverage this information to better understand who is served, their specific needs, and how Sigmund's can better communicate with them.

Market Demographics

The profile for Sigmund's Gourmet Pasta's customer consists of the following geographic, demographic, and behavior factors:

Geographics;

- Our immediate geographic target is the city of Eugene with a population of 130,00.
- A 15 mile geographic area is in need of our services.
- The total targeted population is estimated at 46,000.

Demographics;

- Male and female.
- Ages 25-50, this is the segment that makes up 53% of the Eugene market according to the Eugene Chamber of Commerce.
- Young professionals who work close to the location.
- Yuppies.
- Have attended college and/or graduate school.
- An income over $40,000.
- Eat out several times a week.
- Tend to patronize higher quality restaurants.
- Are cognizant about their health.

Behavior Factors;

- Enjoy a high quality meal without the mess of making it themselves.
- When ordering, health concerns in regard to foods are taken into account.
- There is value attributed to the appearance or presentation of food.

Market Needs

Sigmund's Gourmet Pasta is providing its customers with a wide selection of high-quality pasta dishes and salads that are unique and pleasing in presentation, offer a wide selection of health conscious choices, and utilizes top-shelf ingredients. Sigmund's Gourmet Pasta seeks to fulfill the following benefits that are important to their customers.

- Selection. A wide choice of pasta and salad options.
- Accessibility. The patron can gain access to the restaurant with minimal waits and can choose the option of dine in or take out.
- Customer service. The patron will be impressed with the level of attention that they receive.
- Competitive pricing. All products/services will be competitively priced relative to comparable high-end pasta/Italian restaurants.

Market Trends

The market trend for restaurants is headed toward a more sophisticated customer. The restaurant patron today relative to yesterday is more sophisticated in a number of different ways.

- Food quality. The preference for high-quality ingredients is increasing as customers are learning to appreciate the qualitative differences.
- Presentation/appearance. As presentation of an element of the culinary experience becomes more pervasive, patrons are learning to appreciate this aspect of the industry.
- Health consciousness. As Americans in general are more cognizant of their health, evidenced by the increase in individuals exercising and health club memberships, patrons are requesting more healthy alternatives when they eat out. They recognize that an entree can be quite tasty, and reasonably good for you.

- Selection. People are demanding a larger selection of foods, they are no longer accepting a limited menu.

The reason for this trend is that within the last couple of years the restaurant offerings have increased, providing customers with new choices. Restaurant patrons no longer need to accept a limited number of options. With more choices, patrons have become more sophisticated. This trend is intuitive as you can observe a more sophisticated restaurant patron in larger city markets such as Seattle, Portland or New York where there are more choices available.

Market Growth

In 1999, the global pasta market reached $8 billion dollars. Pasta sales are estimated to grow by at least 10% for the next few years. This growth can be attributed to several different factors. The first factor is an appreciation for health conscious food. While not all pasta is "good for you," particularly the cream based sauces, pasta can be very tasty yet health conscious at the same time. Pasta is seen as a healthy food because of its high percentage of carbohydrates relative to fat.

Another variable that is contributing to market growth is an increase in the number of hours our demographic is working. Over the last five years the number of hours spent at work of our archetype customer has significantly increased. As the number of work hours increases, there is a high correlation of people that eat out at restaurants. This is intuitively explained by the fact that with a limited number of hours available each day, people have less time to prepare their meals and eating out is one way to maximize their time.

SWOT Analysis

The following SWOT analysis captures the key strengths and weaknesses within the company, and describes the opportunities and threats facing Sigmund's Gourmet Pasta.

Strengths

- Strong relationships with vendors that offer high-quality ingredients and fast/frequent delivery schedules.
- Excellent staff who are highly trained and very customer attentive.
- Great retail space that is bright, hip, clean, and located in an upscale mall, suburban neighborhood, or urban retail district.
- High customer loyalty among repeat customers.

- High-quality food offerings that exceed competitors offerings in quality, presentation, and price.

Weaknesses

- Sigmund's name lacks brand equity.
- A limited marketing budget to develop brand awareness.
- The struggle to continually appear to be cutting edge.

Opportunities

- Growing market with a significant percentage of the target market still not aware that Sigmund's Gourmet Pasta exists.
- Increasing sales opportunities in take out business.
- The ability to spread overhead over multiple revenue centers. Sigmund's will be able to spread the management overhead costs among the multiple stores, decreasing the fixed costs per store.

Threats

- Competition from local restaurants that respond to Sigmund's Gourmet Pasta's superior offerings.
- Gourmet pasta restaurant chains found in other markets coming to Eugene.
- A slump in the economy reducing customer's disposable income spent on eating out.

Competition

National Competition

- *Pastabilities:* offers consumers their choice of noodles, sauces, and ingredients, allowing the customer to assemble their dish as they wish. Food quality is average.
- *PastaFresh:* has a limited selection but the dishes are assembled with high-quality ingredients. The price point is high, but the food is quite good.
- *Pasta Works:* offers pasta that is reasonably fresh, reasonably innovative and at a lower price point. The company was sold a few years ago, and consequently the direction of management has been stagnant lately and has resulted in excessive employee turnover.
- *Perfect Pasta:* has medium-priced pasta dishes that use average ingredients, no creativity, and less than average store atmosphere. Sigmund's is not sure how this company has been able to grow in size as their whole product is mediocre at best.

Local Competition

- *Restaurant A:* This is an upscale Italian restaurant that has a limited selection of pasta dishes. Although the selection is limited and pricey, the dishes are quite good.
- *Restaurant B:* An Italian restaurant with a decent pasta selection, however quality is inconsistent.
- *Restaurant C:* An upscale restaurant with a large wine selection and good salads. Everything else is mediocre at best and over-priced. Service can often be poor.

Service Offering

Sigmund's has created gourmet pastas and salads that are differentiated and superior to competitors. Customers can taste the quality and freshness of the product in every bite. The following are characteristics of the product:

1. Sigmund's dough for the pasta is made with Italian semolina flour.
2. Cheeses are all imported.
3. Vegetables are organic and fresh with three shipments a week.
4. Meats are all top-shelf varieties, organic when possible.

At Sigmund's, food is not a product, the experience of dining is a service. Sigmund's prides themselves on providing service that is on par with fine dining. This is accomplished through an extensive training program and only hiring experienced employees.

At a Glance—The Protype Sigmund's Store:

- Location: an upscale mall, suburban neighborhood, or urban retail district.
- Design: bright, hip, and clean.
- Size: 1,200-1,700 square feet.
- Employees: six to seven full time.
- Seating: 35-45.
- Types of transactions: 80% dine in, 20% take out.

Keys to Success

Sigmund's site selection criteria are critical to success. Arthur Johnson, former VP Real Estate, Starbucks, helped us identify the following site selection criteria:

- Daytime and evening populations.
- Shopping patterns.

- Car counts.
- Household income levels.

Critical Issues

Sigmund's Gourmet Pasta is still in the speculative stage as a retail restaurant. Its critical issues are:

- Continue to take a modest fiscal approach; expand at a reasonable rate, not for the sake of expansion in itself, but because it is economically wise to. Continue to build brand awareness which will drive customers to existing stores as well as ease the marketing efforts of future stores.

Marketing Strategy

Sigmund's advertising budget is very limited, so the advertising program is simple. Sigmund's will do direct mail, banner ads, and inserts, with inserts in the Register Guard likely to be the most successful of the campaigns. Lastly, Sigmund's will leverage personal relationships to get an articles of Sigmund's in the Register Guard. Previous friends who have had their restaurant featured in the Register Guard has seen a dramatic increase of sales immediately after the article was published in the RG.

Mission

Sigmund's Gourmet Pasta's mission is to provide the customer the finest pasta meal and dining experience. We exist to attract and maintain customers. When we adhere to this maxim, everything else will fall into place. Our services will exceed the expectations of our customers.

Marketing Objectives

1. Maintain positive, steady, growth each month.
2. Generate at least $40,000 in sales per month.
3. Experience an increase in new customers who are turned into long-term customers.
4. Realize a growth strategy of one store per year.

Financial Objectives

1. A double-digit growth rate for each future year.
2. Reduce the overhead per store through disciplined growth.
3. Continue to decrease the variable costs associated with food production.

Target Marketing

The market can be segmented into three target populations:

- Individuals: people that dine in by themselves.
- Families: a group of people, either friends or a group of nuclear relatives dining together.
- Take out: people that prefer to eat Sigmund's food in their home or at a different location than the actual restaurant.

The Sigmund's customer are hungry individuals between the ages of 25 and 50, making up 53% of Eugene (Eugene Chamber of Commerce). Age is not the most defined demographic of this customer base; all age groups enjoy pasta.

The most defined characteristic of the target market is income. Gourmet pasta stores have been very successful in high rent, mixed-use urban areas, such as Northwest 23rd in Portland. These areas have a large day and night population consisting of business people and families who have household disposable incomes over $40,000.

Combining several key demographic factors, Sigmund's arrives at a profile of the primary customer as follows:

- Sophisticated families who live nearby.
- Young professionals who work close to the location.
- Shoppers who patronize the high rent stores.

Positioning

Sigmund's Gourmet Pasta will position itself as a reasonably priced, upscale, gourmet pasta restaurant. Eugene consumers who appreciate high-quality food will recognize the value and unique offerings of Sigmund's Gourmet Pasta. Patrons will be both single as well as families, ages 25-50. Sigmund's Gourmet Pasta positioning will leverage their competitive edge:

- Product. The product will have the freshest ingredients including homemade pasta, imported cheeses, organic vegetables, and top-shelf meats. The product will also be developed to enhance presentation, everything will be aesthetically pleasing.
- Service. Customer service will be the priority. All employees will ensure that the customers are having the most pleasant dining experience. All employees will go through an extensive training program and only experienced people will be hired.

By offering a superior product, coupled with superior service, Sigmund's will excel relative to the competition.

Strategy Pyramids

The single objective is to position Sigmund's as the premier gourmet pasta restaurant in the Eugene, OR area, commanding a majority of the market share within five years. The marketing strategy will seek to first create customer awareness regarding their services offered, develop that customer base, and work toward building customer loyalty and referrals. The message that Sigmund's will seek to communicate is that Sigmund's offers the freshest, most creative, health conscious, reasonably priced, gourmet pasta in Eugene. This message will be communicated through a variety of methods. The first will be direct mail. The direct mail campaign will be a way to communicate directly with the consumer. Sigmund's will also use banner ads and inserts in the Register Guard. This will be particularly effective because the Register Guard is a popular local paper that is consulted when people are looking for things to do in Eugene. The last method for communicating Sigmund's message is through a grassroots PR campaign.

This campaign will leverage personal relationships with people on staff of the Register Guard to get a couple of articles written about Sigmund's. One will be from the business point of view, talking about the opening of the restaurant and the people behind the venture. This is likely to be run in the business section. The second article will be a food review. In speaking with many different retailers and restauranteurs, significant increases of traffic followed articles in the Register Guard. Because of this level of effectiveness and low/zero cost, Sigmund's will work hard to get press in the Register Guard.

Marketing Mix

Sigmund's marketing mix is comprised of these following approaches to pricing, distribution, advertising and promotion, and customer service.

- Pricing. Sigmund's pricing scheme is that the product cost is 45% of the total retail price.
- Distribution. Sigmund's food will be distributed through a take out model where customers can call in their order and come to the restaurant to pick it up, come into the restaurant, place the take out order and wait for it to be completed, or come in and dine at the restaurant.

- Advertising and Promotion. The most successful advertising will be banner ads and inserts in the Register Guard as well as a PR campaign of informational articles and reviews also within the Register Guard.
- Customer Service. Obsessive customer attention is the mantra. Sigmund's philosophy is that whatever needs to be done to make the customer happy must occur, even at the expense of short-term profits.

  In the long term, this investment will pay off with a fiercly loyal customer base who is extremely vocal to their friends with referrals.

Marketing Research

During the initial phases of the marketing plan development, several focus groups were held to gain insight into a variety of patrons of restaurants. These focus groups provided useful insight into the decisions, and decision making processes, of consumers. An additional source of market research that is dynamic is a feedback mechanism based on a suggestion card system.

The suggestion card system has several statements that patrons are asked to rate in terms of a given scale.

There are also several open ended questions that allow the customer to freely offer constructive criticism or praise. Sigmund's will work hard to implement reasonable suggestions in order to improve their service offerings as well as show their commitment to the customer that their suggestions are valued.

The last source of market research is competitive analysis/ appreciation. Sigmund's will continually patronize local restaurants for two reasons.

The first is for competitive analysis, providing Sigmund's with timely information regarding other restaurant's service offering. The second reason is that local business owners, particularly restaurant owners, are often part of an informal fraternal organization where they support each other's business.

Financials, Budgets, and Forecasts

This section will offer a financial overview of Sigmund's Gourmet Pasta as it relates to the marketing activities. Sigmund's will address Break-even Analysis, sales forecasts, expense forecasts, and how those link to the marketing strategy.

Break-even Analysis

The Break-even Analysis indicates that $40,000 will be needed in monthly revenue to reach the break-even point.

Sales Forecast

The first two months will be used to get the restaurant up and running. By month three things will get busier. Sales will gradually increase with profitability being reached by the beginning of year two.

Expense Forecast

Marketing expenses are to be budgeted so that they are ramped up for months two through four and then lower and plateau from month five to month 10. Restaurants typically have increased business in the fall. This generally occurs because during the summer, when the weather is nice and it does not get dark until late, people tend to eat out less. From months 10-12, the marketing costs will increase again.

Controls

The purpose of Sigmund's Gourmet Pasta's marketing plan is to serve as a guide for the organization.

Implementation Milestones

The following milestones identify the key marketing programs. It is important to accomplish each one on time, and on budget.

Marketing Organization

Kevin Lewis, the owner, is primarily responsible for marketing activities. This is in addition to his other responsibilities, and does depend on some outside resources for graphic design work, and creativity.

Contingency Planning

Difficulties and risks:

- Problems generating visibility.
- Overly aggressive and debilitating actions by competitors.
- An entry into the Eugene market of an already existing, franchised gourmet pasta restaurant.

Worst case risks may include:

- Determining that the business cannot support itself on an ongoing basis.
- Having to liquidate equipment or intellectual property to cover liabilities.

Sample Menus

Lunch

Appetizers and Small Lunches :

- Vegetable Dumplings $7 Six steamed Chinese vegetable dumplings served with two dipping sauces
- Fresh Squid Salad $9 Fresh Squid marinated in garlic, lemons, fresh Italian parsley and extra virgin olive oil
- Gravlax $9 Served with honey mustard sauce, mixed greens and grilled toastets
- Herring Plate $11 Mustard, Garlic, Tomato, Pickled, Dill & Matte Herring served with boiled potato, sour cream, chives and red onion
- Swedish Shrimp Salad $11 Served with lettuce, tomato, yellow asparagus, hard boiled egg in a sweet chill dressing
- Three Crostini's $7 Brushetta, Olive Tapanade and Mushroom & Goat cheese
- Baked Potato $7 Served with Shrimp & Crab Salad or Curry Chicken Salad
- Chicken Wrap $8 Grilled Chicken, carrots, scallion and peanut sauce in a tortilla wrap
- Joe's Sushi Rolls $11 Today's home made Sushi rolls. Ask your waiter.
- Asian Noodles $8 A cup of Asian noodles , served with fresh edamame and spinach in a sweet Thai chili dressing
- Tuna Nicoise Ficelles $7 Fresh chunks of tuna, egg slices, tomatoes, red leaf lettuce and black olives.

Soup

- Split Pea Soup $5
- Cucumber and Avocado Soup $6 With tomato & basil salad
- Fish Soup small $7 large $9 With saffron, salmon, fresh mussels, claims and lobster meat

Salads

- Small Mixed Green Salad $6 With house dressing
- Spinach, mushrooms, red onion, yellow peppers and smoked bacon $9 Served with a warm mustard - bacon vinaigrette

- Thai Chicken Salad $11 Tomato, green pepper, cucumber, lettuce and white rice in a spicy sweet Thai dressing
- West Coast Salad $12 Shrimp, crab meat, mussels, green peas, yellow asparagus, tomatoes, mushrooms, green leaf and iceberg lettuce
- Flank Steak Salad $11 Thin slices of flank steak marinated in an Asian dressing and served with mesclun salad and fresh mango
- Gravlax Salad $12 Thin slices of gravlax, mesclun salad, cucumber and sweet mustard dressing
- Vegetarian Salad $11 Spinach, mushrooms, roasted peppers, carrots, eggplant, squash topped with olives and served warm with a lemon - tomato vinaigrette.

Today's Lunch Special $13 Including a Small Salad and Fresh Baked Bread

- Meat: Red Curry Chicken With Basmati rice and fresh steamed broccoli or
- Fish: Grilled Honey Mustard Salmon With red roasted potatoes or
- Pasta: Linguine with fresh pesto sauce.

Today's Business Lunch

Appetizer

- Gravlax or
- Prosciutto Salad.

Entree

- Short Ribs of Beef Served with mango chutney and mashed sweet potatoes with a port wine sauce or
- Dill Crusted Cod Served with roasted potatoes, carrots and citrus broth or
- Swedish Meatballs Served with mashed potatoes, lingon berries and pickled cucumbers.

Dessert

- Coconut Rice Pudding With mango sorbet or
- Chocolate Cake With whipped cream and Raspberry sauce.

Other Lunch Specialties

These menu-items will change every week;

- Smoked Salmon Panini $9 With a Swedish Remoulade
- Tuna and Salmon Burger $12 Served on a sourdough roll with Pineapple Chutney
- Seared Halibut with fresh Spring vegetables $15 Served with a Tomato - Cucumber Feta Salad
- Italian Sausage and Wild Mushroom Risotto $13
- Home made Butternut Squash Ravioli's $13 Served in a white wine sauce with fresh spinach
- Pan-Roasted Chicken with Cognac-Peppercorn Sauce $15 Served with fresh broccoli and mashed potatoes
- Spiced Pork Tenderloin with Pineapple-Avocado Salsa $15 Served with black beans and rice
- Asian Vegetables with Tofu and Coconut milk $13 Fresh vegetables with tofu and coconut milk
- Turkey Burger with Chipotle-Chili Tartar Sauce $12 Served on a bun with arugula, red onion and plum tomatoes
- Plate $13 An assortment of Swedish Delicatesis.

Sandwiches

- Shrimp (open face) $7 Swedish Shrimp, tomato, hard boiled egg, cucumber, lemon aioli and topped with fresh dill, served on white bread
- Meatballs (open face) $7 Swedish Meatballs and pickled beat salad, served on pumpernickel bread
- Turkey Wrap $6 Fresh sliced honey turkey, green lettuce, roasted peppers, sprouts and feta cheese wrapped in a tortilla
- Egg Salad & Tuna Salad $6 Two classics!
- Warm Ham and Pineapple $7 Served with cheese, mixed greens and sweet chili dressing
- Tomato and Mozzarella $6 Served in French Baguette.

Hors d'oeuvre Platters A variety of appetizers;

- For two people $10
- For four people $20
- For six people $30
- For eight people $40.

Dinner

Appetizers

- Small Mixed Green Salad $6 Served with house dressing
- Herring Plate $11 Mustard, Garlic, Tomato, Pickled, Dill & Matte Herring served with boiled potato, sour cream, chives and red onion
- Gravlax $9 Served with honey mustard sauce, mixed greens and grilled toastets
- Plate $15 for two people An assortment of Swedish Delicatesis
- Joe's Sushi Rolls $11 Today's home made sushi rolls. Ask your waiter.
- Mushroom Sauta Torte $7 Served warm with goat cheese
- Oysters with spicy jicama salad $11 Six oysters served on the half shell with a lemon sweet vinaigrette
- Brushetta a la Jacob $7 Chopped tomato, garlic ad fresh basil served on crostini
- Fresh Squid Salad $11 Fresh Squid marinated in garlic, lemons, fresh Italian parsley and extra virgin olive oil
- Spicy Fresh Tuna Tartar $9 Fresh Tuna rolled in cucumber and topped with a spicy chili sesame seed dressing
- Chicken Skewers $8 Served with a tropical peanut dipping sauce
- Prosciutto and Avocado Salad $8 Served with Frisee and Arugula Salad in a lemon vinaigrette.

Entrees

Fish

- Fish Soup $9 With saffron, salmon, fresh mussels, claims and lobster meat
- Sea Bass with Cous Cous Risotto $13 Sauteed Sea Bass with cous risotto, green peas and fresh corn
- Roasted Cod with Spinach and Mushrooms $14 Served with red roasted potatoes
- Filet of Mackerel $14 Grilled filet of Mackerel, served with a beet and potato roulade
- Grilled Atlantic Salmon $14 Served with sauteed onions, leeks, artichokes and potatoes in a lemon jus

- Shrimp Charmoula $16 Moroccan Shrimp Skewers marinated in an aromatic herb-and-oil mixture before grilling
- Ahi Tuna with white beans and sundried tomato sauce $15 Seared Ahi Tuna with white beans and a fresh sundried tomato sauce
- Mussel, Clams and Shrimp with saffron risotto $17 Fresh mussels, clams and shrimp in a saffron risotto.

Pasta

- Linda's Pad Thai $12 Pad Thai noodles, garlic, bean sprouts, red and yellow peppers, eggs, leeks, peanuts, tofu in a tamarind - fish sauce. Add shrimp: $3 add chicken: $2
- Angel hair pasta with fresh roma tomatoes, garlic and fresh basil $12 Served in a light white wine sauce with crushed red peppers
- Vegetarian Lasagna $12
- Home made Butternut Squash Ravioli's $14 Served in a white wine sauce with fresh spinach and roma tomatoes.

Meat

- Swedish Meatballs $13 With mashed potatoes, lingon berries and pickled cucumbers
- Chicken Thai $13 Sauteed Chicken Breast pieces in a spicy Thai sauce. Served with jasmine rice.
- Mango Chutney Filet Tenderloin $15 Filet Tenderloins with mango chutney and steamed broccoli
- Duck Breast with orange, honey and champagne sauce $16 Seared duck breast with a sweet sizzling champagne sauce. Served with white beans.
- Rosemary and Garlic Lamb Chops $17 Grilled Lamb chops with red roasted potatoes and red wine sauce
- Pepper Steak $18 Beef Tenderloin with black pepper and cognac sauce. Served with fried potatoes.
- Tornadoes with Bearnie Sauce and Red Wine Sauce $19 Two beef tornadoes, bearnie sauce and red wine sauce. Served with a baked potato.
- Barbecue Beef Ribs $15 Served with baked beans and fried potatoes.

Chapter 5

Beverage Industry

The food industry is on a roller coaster ride as Indians continue to have a feast. Fuelled by– large disposable incomes – the food sector is witnessing a remarkable change in consumption patterns, especially in terms of food.

Food retailing has come of age — from a period when food items were sold in small road side grocer shops & mandis, haats and bazzars by vendors to a stage when food products (processed and groceries) are retailed through supermarket stores where consumers can inspect, select and pick up the products they like in a comfortable ambience and still pay a fair price for the product and the merchandise and sometimes even pay less than the price they would have paid at the nearest food stores. Shopping for groceries is no longer a strenuous and uncomfortable affair. Instead, it is a pleasurable experience. From simple trading activity, food retailing is now heading to the status of an industry. Imagine yourself walking through the air conditioned lanes smelling fresh food and groceries, enjoying light music, experiencing five star ambiences and above all buying food products (vegetables, spices and beverages) without bargaining with the vendors.

Food retail has surpassed the dominating apparel and accessories sector. "Contrary to the belief that fashion is the largest segment of organised retail in India, food & beverages is the major segment, worth Rs. 8,97,000 crore," said Arvind Singhal, Chairman, Technopak, at the Indian Retail Forum held in Mumbai. There is an old industry saying that " customer is king" Food retailers today will update that saying to "customers is the dictator" because of fierce competition, new technology and business practices the market power of customer is strong and growing stronger.

The Food Retail Tale

Traditionally, Indians were used to buying their sugar, wheat, pulses, rice etc. from their neighborhood baniya. The majority of food and food products were and still are retailed through neighborhood kirana (baniya) stores. A typical kirana store has a retail area of 200 sq. ft. and sells 500 to 800 stock keeping units (SKUs).

The kirana stores focus on dry food products because the infrastructure for cold storage is lacking. The majority of fresh produce is sold from the carts of travelling vendors. Such produce is deemed to be of low product quality, variety and hygiene. This concept is still popular in B-class and II Tier cities of India and giving employment to thousands of its inhabitants.

Initially the food retail format was seen in A-class cities like Mumbai, Delhi, Chennai which had co-operative stores like "Apna Bazaar" in Mumbai and "Kendriya Bhandar" in Delhi. Both were very successful and are operating many outlets in all strategic localities in the city.

Escorts group in the late eighties diversified into non-auto sectors by getting into agri business or food business. It came out with first "Nanz" store at South Extension in Delhi in 1990.

The first visible sign of the change in food retailing was seen in mid-eighties. Around that time a few new food stores were set up in all metro cities in India. Calcutta was the only exception where it started a little later. At that time couple of leading food stores started operating such as "Morning Stores" and "Modern Stores" in Delhi, "Nilgiri" in Bangalore, and "Food Land" in Mumbai "Spencers Food Stores" in Chennai.

Spencers were the first to tie up with a Singapore based large retail chain — "Dairyland" and had set up the food stores in Chennai. This was a technology tie up. In Mumbai, Garware group during the late eighties had set up a large food store, which is now reported to have been closed down.

Until the late 1990s, food retailing has been concentrated in the south of the country. Southern India has been witnessing revolutionized activity in food retailing.

It has thoroughly experienced the food retailing in various formats such as the supermarkets, hypermarkets and neighbourhood stores. These include Food World, Subhiksha, Nilgiris, Margin Free, and Big Bazaar.

The reason being that most entrepreneurs who started organised retail came from southern India and the cost of real estate in the southern region was less than other regions (particularly north and west). Since then, however, organised food retailing has emerged across the country, inspired by the presence of high potential markets in the north, west and east as well as the success of some non-food retailers and food services companies in these regions.

Neighbourhood Stores

In India about 90% of food purchases are made within a distance of 1.5 km from the customer's home. This means that an organised retailer would need to have a 'neighbourhood store' close to customers in order to capture the share of wallet that is spent on food.

These stores would cater to the consumer's daily and weekly needs. The outlets closest to a neighbourhood store in India are 'Safal' outlets operated by Mother Dairy in Delhi, Margin Free in Kerala and Subhiksha.

Food Research Institutions in India

Produce Little Value

Food Science and Technology Education in India was initially masterminded by pure scientists in the field of chemistry, applied chemistry and biochemistry. The boderline between the basic science and technology and engineering was therefore not clearly demarcated at the initial stage.

As a result, research projects that were undertaken in most of our premium food research institutions including Central Food Technological Research Institute (CFTRI) were designed on the line of basic science exploration in the areas of biochemistry, protein chemistry, nutrition and food processing. Those who have ventured into the processing and technology also focussed only on physiochemical changes of food during processing and process parameters optimization.

Those who have ventured a little ahead thus incorporated process conditions standardisations to improve product quality. Nothing therefore happened in terms of any major breakthrough in process technology in India.

Our food research institutes including universities did very little in terms of technology upgradation and new product development which are of any commercial significance. Whatever little we have seen in new product development and product quality upgradation is the result of the efforts on the part of the industry largely triggered from the need as a survival strategy in the face of competition.

Some of these research institutions also suffer from the syndrome of reinventing the wheel. Crores of rupees have been spent by young food researchers reproducing the knowledge which are either known and or explored elsewhere. This therefore has put the industry, in particular and the nation, in general into the back gear.

We have to painfully observe in various conferences and seminars that the research findings are presented with lot of gusto which are nothing more than the age old findings and established processes.

Another factor that has marked the failure of food research activities in India to create any dent in terms of value addition in the processing and technology is due to the fact that there was little or no collaborative efforts between industry and research institutions.

The mutual mistrust and lack of confidence on the part of the industry leaders on the ability of the institutional research to deliver value has increased with the passage of time. As a result, researchers have ventured into their own chosen subjects while industries have decided to go it alone.

Formation of an independent Ministry of Food Processing Industry which has now been reduced to the status of a department did not help change this scenario. The result is inevitable. While we can see a sea change in other industry categories in terms of product quality and value addition, in case of processed food industry changes are miniscule to say the least, and that too largely due to the efforts of the industry. On the top of it, some of our premium institutions are in reality in a state of decay in terms of quality of research and education.

The research activities that have been initiated earlier by the pioneers in food research and education have lost the inertia in the later years and thus died down. In this context we can remember the name of Prof. V. Subramaniam, first Director of CFTRI, Prof. A.N. Bose, first Director of Central Marine Product Research Institute in Cochin, Prof. A. Sreenivasan-First Director of Food Technology Dept. of BARC. Prof. Gopalan, first Director of Nutrition Research Institute (NII) at Hyderabad etc. CFTRI has seen some of the stalwarts in later years which include the names like Mr. C.P. Natarajan, Dr. H.A.B. Parpia, Dr. B.L. Amia and few others.

National Dairy Research Institute (NDRI) in my reckoning, do have a creditable record of quality research output and in that also contribution of Dr. K.K. Iya and most importantly that of Dr. N. Ganguly are noteworthy.

In oil technology Dr. K.T. Acharya's and Prof. M.M. Chakraborti's contribution is also significant. In the areas of biotechnology and biochemical engineering Prof. T.K. Ghose's contribution is pioneering.

He has, in fact, given the shape to biochemical engineering education and research in India and was associated with Jadavpur University, HBTI-Kanpur and IIT-Delhi. Defece Food Research Laboratory (DFRL) in Mysore also worked on projects of specific interest to our defence forces.

In the areas of research and education we now have a generation gap. After the retirement and or passing away of some of these above mentioned stalwarts most of these food research institutions have been reduced to insignificance. If I describe some of those as "in a pathetic state of affairs" it will not be any exaggeration and in that list my old institution namely, Jadavpur University also features. The reasons of such state of affairs are many. One, – relatively inferior quality of people have come into the research and education, two, – the system has made them dull and non motivated, and most importantly, our stalwarts have not taken any pains to leave any legacy to continue.

As a result, with their disappearance from the scene the things have drifted and ultimately came to a grinding halt. And that is a major failure.

There are enthusiastic scientists like Prof. A. Sreenivasan who developed protein food by name 'Spert' – a product manufactured and marketed by Sandoz. Prof. Sreenivasan also developed the process for production of egg powder at Foods and Inns where he was a consultant.

Foods and Inns share still is quoted at Rs.60 today and the company is still doing the same thing that they used to do during Prof. A. Sreenivasan's time. When 'Spert' was launched during the late seventies I was discussing, one evening, with Prof. Sreenivasan at his Gautam Apartments flat in Bandra in Mumbai and asked him whether he really believed that there was a need of any protein food for those who can afford to buy 'Spert'. He confided in me and told me "Look Rajat if one takes full meals of only dal and rice one's proteins requirements is met and these are all humbug".

I responded by saying that "then the consumers for these so called protein foods are really taken for a ride." The marketers like us know that if there is no permanent value proposition, products do not have a long term prospect in the marketplace.

Possibly, for this reason protein foods gradually disappeared from the market like fads. Of late, I see the decline of the sales of Complain in the hands of Heinz. Prof. Sreenivasan changed jobs in public sector research institutions like one changes or sometimes forced to change in private sector.

He used to be a man of very strong conviction and a forceful character and as a result nobody liked him. But his contribution to the industry is significant.

Prof. Bose was a strong disciplinarian and perfectionist. His contribution in food research and education in noteworthy. He used to come to the department in Jadavpur University exactly at ten in the morning and used to take a round of the laboratories and classrooms.

During his time a lot of research projects were there in the department which was in the global map with respect to food research and education. Today one cannot find a teacher in the department till twelve noon.

No one comes before that. Jadavpur University did pioneering work in rice process technology, marine products processing and protein technology etc. in earlier years.

As it stands now the future development thrust for the processed food industry therefore largely has to be provided by the private sector industries itself. As the market is becoming increasingly competitive the innovations in terms of quality, cost, packaging and product variants and new products will be the name of the game.

And private sectors will have to provide the support for their own survival. This is possible. For example, Pepsi Cola in USA has kept on their roll over a dozen very senior level scientists specialised in each area of their activities to support the company's competitive strategy. This team of specialists are basically the technical backbone working for various ongoing and existing and future projects of Pepsi. In that team Dr. Shanti R. Bhowmik, a close friend of mine and a former director of CFTRI is there. Dr. Bhowmik spent 2-yr. time at CFTRI and having got fed up of the bureaucracy and the inner politics of the institution and poor level of research activities decided to return to USA again.

This was narrated to me by Dr. Bhowmik. Of course, he has also praised some of the research scientists for their quality of work and dedication and in that list the name of present Director of CFTRI Dr. V. Prakash's name featured.

I wonder if India has to be the leader in this new millennium, technical competency has to be improved for our domestic industry so that we can have more Tata-Tetly kind of organisation being created in global scenario. And that can happen only with private sector initiatives. Let us not keep much hope on our institutional and Govt.. run organisation's research effort. That has not happened in the past and I don't see that is happening in future either. The reality is that past of food research in India was more glorious than what it is now.

Acquisition of Domestic Food Business by MNCs

Liberalisation has brought in through the entry of large multinational and transnational corporations, foreign investment in this sector. This resulted in competition, technological upgradation and market expansion. In the face of the competition, domestic Industries are gradually losing market share and thus selling their businesses to the new entrant MNCs well before the value of the brand and businesses drops further due to ongoing onslaught of multinational brands.

This is happening because MNCs have much greater resources to put behind their brand and business and also have long term vision and sustainability. Domestic industries are no comparison. The first attack came in soft beverages industry with Coke acquiring Parle brands and Pepsi acquiring Dukes and with the entry of Cadbury Schweepes in this sector entire soft drink industry is now in control of MNCs.

MNCs such as Nestle, Brooke Bond and SmithKline & Beechern controlled earlier only hot beverages (both white and brown segments). But soft beverage sector was dominated by domestic industries. With Coke and Pepsi acquiring Indian businesses, total nonalcoholic beverage industry is now under control of MNCs.

Recently Sara Lee — a US multinational has acquired Nutrine — a large Indian bakery and confectionery business. Britannia is already under control of Danone group of France. Only large bakery and confectionery left to be acquired in private sector is Parle Products.

And with the entries of confectionery giants like Perfetti, Wriggly and Agrilimen (a Spanish confectioner in joint venture with Dabur but in the process of acquiring Dabur's equity to have 100% control) confectionery and bakery sector, will also become the domain of MNCs.

With Nestles' strong presence in Tomato Ketchup and Pickles and Hindustan Lever acquiring Kissan and Dipy brands from UB group through Brooke Bond route fruit and vegetable sector is also dominated by multinationals.

Heinz, the world's largest Tomato Ketchup manufacturer had acquired food business of Glaxo. Due to strategic restructuring of portfolio Glaxo decided to divest their dairy business with established brands like Complain, Glucon-D and Farex. Hindustan Lever, on the other hand, through a strategic alliance route, took control of Kwality Ice-cream as well as 100% Ice-cream brand of Jagatjit.

Hindustan Lever with their own presence in dairy sectors, and with the introduction of Baskin Robbins and Walls has emerged as a dominant player in dairy segment, in general and Ice-cream segment, in particular. Hindustan Lever (HLL) also acquired Dollop Ice-cream brand from Cadburys'.

HLL, however, recently announced their intention to sell-off their Dairy unit in UP. Ice-cream industry, which is reserved for small-scale sector, has now been in the control of multinationals through a different route. The argument that ice-cream is actually a frozen dessert — a nomenclature that has been accepted by both food and law ministry of Govt. of India — paved the way for Hindustan Lever for this strategic acquisition.

The earlier successful domestic businesses such as Dalmia Dairy, Foremost Industry and Jagatjit Industry in dairy sector are all in deep trouble — and MNCs have taken a control of this sector as well.

National Dairy Development Board (NDDB) with its strong brand Amul and with its deep root extended up to farmers level as a means of backward integration will of course, remains a significant force to reckon with. But nobody will deny that in branded dairy products Amul has to put up a constant fight against multinational brands.

Vadilal is also a significant player in ice-cream but their presence is limited in Western India and for how long they will remain has to be seen. Kraft Foods a unit of Phillips Morris is holding a licence to set up 100% owned entity and they are expected to come and set up their operation anytime.

In the convenience foods — 'Corn Products', a US multinational is fast making progress. This is an old company, which did not make any significant contribution in the Indian food business largely because of lack of interest from their principals.

CPC-International is now fast making progress with their well known Rex and Brown & Poison and Knorr soup brands. Recently they announced the acquisition of DCW Chemicals — makers of Captain Cook brands, which ran into rough weather because of the objection taken by some miller in Haryana to whom allegedly DCW. has contractual obligations.

Shortly, we will also have products in joint venture with Paoma Industries, the manufacturer of Rasna soft drink mixes and Campbell a world leader in canned foods. Godrej, on the other hand, has joined hands with Pilsburry a leading player in the convenience foods business to boost their business activities. Filsburry has already introduced instant Cake mix.

In this category others including Gits brand are relatively smaller and regional players. Apparently, first change that we have witnessed is of acquisitions of domestic brands in frontiers of food business by MNCs and alongwith although at a much lower scale a few joint ventures. Even the public sector undertaking, Modern Foods although making profit is up for sale in line with the Govt. policy of PSU disinvestment and ANZ Bank has been appointed to find a buyer for the same.

The first lesson that we have learnt as the effect of liberalisation is acquisition of successful domestic food business by MNCs.

Global Competition

Ten years ago Indian processed food industries were crying about high taxation, lack of infrastructural support, nonavailability of quality raw material at competitive price, nonavailability of latest packaging machinery and packing material etc. as the major reasons for poor or sluggish growth of this sector. But the industry was still growing then, albeit not at the expected level. Competition was also there amongst domestic processors with their MNC counterparts. Because, large MNC food companies like Nestle, SmithKline Becheern, Britannia, Glaxo, Hindustan Lever etc. are there for ages.

But domestic industries were not worried about their presence. There was healthy competition. In some areas domestic industries were doing even better than some of their nearest competitors in multinationals. What went wrong then? Why domestic industries are now not able to fight multinational giants? This is because the rules of the game have changed. Post liberalisation, we have now entered into an era of survival of the fittest.

Only couple of years ago we had only Amul Cheese. Britannia first successfully introduced Cheese from cow milk followed by the introduction of French cheese 'Le'-Bon" from Dabur-Le'-Bon joint venture.

Philips Morris — owner of Kraft Foods — is sitting waiting a formal licence to set up 100% owned company in India. They carried out lot of research for introduction of Kraft cheese in India. Their project now has been put in cold storage and they are still waiting and watching the market for processed food industry.

Upto mid eighties Kissan was the number one brand in Tomato Ketchup and around that time the competition came from Maggie — a Nestle's brand. With Maggie launching several varieties of Tomato Ketchup there was a growth in the market. With Kissan and Maggie fighting neck to neck other smaller but established brands like Dipy's from Herbertsons, Volfruit from Voltas, Noga of Nagpur Orange Federation and SunSip of Wimco gradually disappeared from the market. Delmonte now has joined hands with Paoma Industries, the manufacturers of Rasna Brands of soft drink concentrate and is expected to launch Tomato Ketchup shortly. The biggest global player in Tomato Ketchup Heinz is still waiting and watching. It was expected that they will launch Tomato Ketchup in Indian market soon. Some of the unknown local brands of course still existing only on price competition.

The price competition has also kept some of the other categories of food products still alive. For example, the age old Corn Flakes from Mohan Meakin under the brand name Champion still sells almost the same quantity (reportedly 500 MT per annum) against the much better quality Kellogs brand which sells around 300 MT per year with all its variants.

Till today we have only Parle and Britannia as two major brands in biscuits. Kellog is entering that segment gradually. Cadbury and SmithKline Becheem have nominal presence with speciality biscuits.

Shortly, we are going to see more cookies and speciality biscuits from Sara Lee, which has acquired Nutrine.

In the snack foods category we have two successful Indian players namely Uncle Chips brand of Amrit Foods and Pepi of SM Dye Chem, others were local such as Fun Flip, Crax etc. Soon entered Pepsi with their Ruffles brand and soon extended to Indian snacks by introducing bhujjias which Bikanerwala makes for them to market.

In confectionery we have over half a dozen global players struggling here for brand share and fighting against Indian players like National, Norton, Parle, Parrys etc. Amongst MNCs in this field leading are Perfetti, Wriggley, Agrilimen, Park Davis (with their chicklet brand).

Competition is also seen in organised food retail chains. Couple of years ago we had only Narula running such an organised retail food chain. Today we have everybody namely. McDonald, Kentucky's Fried Chicken, Dorninos, Pizza Hut of Pepsi's and many more. In some of these categories domestic brands are still a leader-for example, Uncle Chips, Frooty, Rasna etc. But how long they will be able to retain their leadership is a big question mark? If, potato price continues to remain at the present level possibly Pepsi's Raffle will still continue even after incurring losses but Uncle Chip will have to close its shutters.

As others are doing — for example Billy's Ltd. — an Indian company in technical assistance with Florigo of Netherlands had successfully introduced potato chips in Eastern India and was reportedly doing well. As potato price started shooting up the company could not afford to sell its brand at a competitive price and is now in the process of closing down. With highly fluctuating input price, domestic industries will be all the more vulnerable like the one as we have seen in case of Billy's.

The second lesson that we have learnt is that in each category now we have global competition with more players expected to jump into the fray. Domestic processors will have to learn to survive competition. But consumers are the real beneficiaries, they have now a wide choice.

Technological Upgradation Leading to Superior Quality

Food processing industry in India has undergone a total metamorphosis over the last six to seven years, in terms of types, variety, quality and presentation which is a result of liberalisation leading to foreign direct investment in frontiers of processed food sectors. Multinational and transnational corporations made significant contribution.

This was possible due to market significant technological upgradation in this sector. It is not that the latest state-of-the-art technologies in this sector were not known to us. On the contrary, we had the fullest capability and technical know-how.

In seventies and eighties we used to talk about in seminars and conferences the new types of packaging, products and processes but industries never felt the necessity to upgrade or invest in new technology. Nor did they have the resources to support such programme.

Reason was is the apprehension all around that there would be no market for such products. And we continued to produce products that are of traditional nature using age-old technologies. Besides, there would were licensing restrictions, reservation of many categories under small-scale sectors etc., resulting of into poor or negligible upgradation of the industry. Things have changed since then.

With the technological upgradation we now see an all round improvement in processed food products quality. The competition has forced the processors to improve and the upgrade the quality. Earlier we had limited categories, for example, jam, jellies, marmalades, squashes, cordials, ketchup and sauces in fruit and vegetable sector, ice cream, spray dried milk powder, condensed milk, butter and ghee in dairy sector, bread, biscuits and pancakes in bakery sector, hard boiled confectioneries, one brand and type of chewing gum and limited variety of chocolate in confectionery sector. Today we have not only better quality of these but also a lot of variants and product types and choices. There is also a significant improvement in packaging quality. In seventies we had only glass and cans as packaging materials. In eighties we had seen laminates and other plastic materials coming in.

Today we have the host of packaging types being used to market processed food products. Earlier we used to get fruit juices and ready-to-service (RTS) beverages only in glass bottles. Today we have tetra pack and also asceptic carton pack. Similarly, tomato paste earlier used to be sold in metal cans and the market for this was also very small. Today tomato paste is sold by Godrej Foods in asceptic consumer pack. In packaging of cheese and other dairy products considerable improvement in the product and packaging quality are evidenced. Cheese, as a product was not known to the Indian consumer in seventies and eighties. Only cheese that we had was cottage cheese.

Processed cheese was not there, not to speak of varieties of cheese as they are found and consumed in other countries. I remember during early eighties we had to develop a particular party near Delhi to supply Mozzrella cheese required for pizza to be sold through Mcdowells Fast Food chain which subsequently closed down.

Today although we don't have all varieties of cheese but we have seen a beginning. Britannia and La-bon cheese are definitely gaining ground. And we expect that when Kraft will enter this market there will be a revolution in processed cheese sector in India.

Kraft International has done considerable research on the possibility of producing world class Kraft cheese for Indian market and they came to the conclusion that the milk quality in India is not good enough to produce cheese of Kraft grade. Besides, Kraft cheese is produced from cow milk where as in India bulk of our milk is from buffalow. They even thought of importing the compressed casein block to Nepal under refrigerated condition and set up processing plant there for Indian market in order to take the custom duty advantage because quality is what they can not sacrifice.

But that also was not found to be viable. I told Mr. Puno who is in charge of South East Asian region including Indian subcontinent that starting early is the key. I advised him to start the operation first and cow milk can be made available or can be organised.

But they had still the apprehensions. Britannia subsequently introduced cheese and their sole claim is "Cheese from cow milk" and has done very well if initial responses are any indication. And it is everybody's knowledge that Britannia is not manufacturing it themselves but it is being manufactured in a local plant near Pune under a subcontract manufacturing arrangement.

Britannia is only marketing it. Although process technology support might come from their respective French collaborators namely BSN and LA-BON, but this only proves my contention that technology is known to us but we did not put it to application to tap new and emerging market. Competition is forcing the marketeers to look for better products and new products, which is a direct outcome of the liberalisation policy.

Even the traditional item like Indian snacks (bhujjia) and Bengali sweets manufacturers are also thinking to upgrade their manufacturing technology and product quality. Mr. Agarwal, Managing Director of Bikanerwala who supplies bhujjia to Pepsi Foods recently invited me to visit his plant at Lawrence Road in Delhi and said Pepsi's global chief also visited his plant and was highly impressed to see their process.

When asked what exactly the improvement they have made in their age-old labour oriented process, Mr. Agarwal commented that the packaging of Bujjia is now fully mechanised.

Similarly sweets are also a serious trade worth Rs.16500 crore and in West Bengal alone the market is estimated to be Rs.1000 crores. The Sweet Traders Association in West Bengal recently held a seminar in Calcutta to devise ways of making the sweet making process more hygienic and to ensure optimum use of technology _ said Mr. Robin Pal, General Secretary of the Association. The implications of liberalisation has also reached these so-called traditional foods and the traders now are eyeing for the global market.

Under liberalisation regime consumer is the king. They are getting better value for money and they have now a wide choice to select from.

Proliferation and Extension of Food Brands

Food items are primarily processed to extend its shelf life and to preserve its nutritional value. The secondary purpose of processing of food is to deliver convenience, taste and appeal to the target consumers. Depending upon what consumer segments we are targeting the processing methods, type of food and its taste and packaging and total presentation varies widely. In a country like India where local taste preferences differ from region to region it poses a much bigger problem to produce food products which can have a much larger acceptance and hence a bigger market potential.

But there are some food categories and types for which there is a universal acceptance and hitherto Indian processed food industry only were processing those generic category of foods for all India market. Even new entrants in the processed food industry were only adding brands in the traditional sectors. We have witnessed this phenomenon before liberalisation in different categories. And as a result, mortality rate of processed food brands were quite high. For example, when Amul came into the market with their Amul Butter-Polsons butter disappeared from the scene.

And the successful brand went on extending it to other categories. Having tasted success, Amul brand has been extended across a range of items including cheese, cheese spread, shrikhand, chocolate and ice-cream. Creativity in food industry was, to say the least, very low. There were of course small local companies who attempted to create products for catering to the local taste preferences but the market was so small that the products did not survive the test of time. For example Kissan launched "Bisebelle Bhath" a South Indian delicacy in Southern market in mid eighties only to withdraw a year later.

We at Kissans did considerable work and developed instant rice but market research indicated that the market for instant rice would be insignificant and thus we had to drop the idea.

The same instant rice was used to formulate "Bisebelle bhath". Phillips Morris who got clearance recently for setting up 100% owned entities in India has announced their plans to introduce the instant rice under their global and popular brand "Minute Rice".

And still there are others who tried to create universal appeal and introduced mass market product and have demonstrated big success stories. The example is Rasna from Pioma Industries. But how they did it is a different story and that will be covered in another article.

Yet others have failed in such attempt. In mid eighties steel merchant of Mumbai Mr. Ravi Kiran Agarwal in a joint venture with Kwality Frozen Foods had set up Tasty Bite Eatables Ltd.-a company to get into frozen foods and North Indian and South Indian vegetarian and non-vegetarian ready-to-serve foods.

The concept was-warm and serve. A modern plant was set up near Pune. The products including Mutton curry etc. in retortable pouch were introduced but it did not click. A couple of years later they faced serious financial problems and subsequently the company was sold to Hindustan Lever-who tried hard to establish "Green Valley" frozen foods. It was also later on withdrawn.

Welga Foods in Badaun near Bareilly similarly had to close down their frozen foods plant. Tarai Foods who were also contract manufacturers for Hindustan Lever is still struggling to exist. The only long term player in frozen foods particularly frozen vegetables is "Mafco" — a co-operative in Maharashtra. Food products that were in the high price range and were well ahead of their time did not succeed in India. And thus casualties in certain category were very high.

Before nineties the food industry in India was struggling to establish only the traditional food brands and there was very little or no effort to introduce new foods.

In post liberalisation era we have host of new MNCs with superior marketing skill and bigger resource to their command have extended their brands to a wide variety of local foods and their variants including the areas which were protected niches of small local players. For example, Indian traditional snacks were only local brands even a couple of years ago. But there we have now even global brands. Kissan under UB-group used to sell and market only fruit and vegetable products. Today Kissan brand has been extended to even branded grocery items which were earlier the domain of small local players.

Pepsi's primary business is in soft drinks but in India they have gone into branding of traditional snack foods. Similarly Nestle-a swiss

Food major-has introduced pickles in Indian market. In the traditional food items local companies should be able to add better value and hence can score higher in the market place.

In seventies, 'Bisleri' was the only mineral water which had national presence and the sale was to the tune of approximately one hundred thousand cases valued at about Rs.60 lacs. Today mineral water market is said to be of Rs.200 crores and over hundred brands of mineral water are there.

Mr. Ramesh Chauhan of Parles thinks that mineral water market in next five years time will become Rs.2000 crores and based on that he has drawn up ambitious plan for his Bisleri mineral water.

The competition in mineral water is so high that Mr. Ramesh Chauhan's own brother Mr. Prakash Chauhan who owns "Frooty" brand also introduced a mineral water under the brand name "Bailey". Of late we have multinationals entering the mineral water scene. "Evan" is recently launched which is a well known international brand.

Earlier we did not have any fruit juice in the market. Only canned juice and ready-to-serve beverages in bottles were available. Today we have number of brands in carton and tetrapack. Such as 'Onjus' of Enkay Texofood, 'Real juice' of Dabur, 'Jumpy' of Godrej etc. all packed in convenient asceptic cartons. We will be shortly getting more brands of fruit juice including 'Tropicana' from Seagram.

The proliferation of brands are also seen in branded Grocery products. A few years ago we had 'Tata-salt' as the only refined iodised salt selling in the market and today we have 'Captain Cook' from DCW, 'Kissan Annapurna' from Kissan Products and Catch (with dispenser) from Hi-Tech Foods and more brands to come in.

In some areas such as meat and poultry products branding, packaging, processing and distribution on a national scale has yet to be witnessed. But in many of the processed food categories, as we can see that the existing brand is either extended to a wide variety of foods such as Britannia to Cheese and Long Life Milk in tetrapack etc. or there are proliferation of brands.

In this process, ultimate beneficiary is the consumer as they can expect to get the best value for their money. Competitive pressure will make that possible.

MNCs' Entering Traditional Foods Sectors

Mckensey did a study on the opportunity in India for food and agri business at the behest of Confederation of Indian Industries (CII).

And as expected they projected a huge potential in food sectors and said that the real potential lies in traditional, local food and grocery items — but not so much in westernised processed food. Some of our big multinationals seem to have swallowed the Mckensey advise-pill and got into this sunrise sectors.

Historically, traditional food items were the domain of small local players. Grocery items earlier were sold loose and unbranded. Then some progressive traders started the cleaning, grading and primary packaging. And the next stage of development was branding of primary foods such as Atta (wheat flour), Rice etc. but those were mainly regional brands. Popular among these were "Rose" and "Shakti Bhog" brands in northern India.

Whereas in traditional food items such as spice and pickles there were host of local brands with regional strong-holds such as 'Cookme' spice powder in East, 'Bedekar' pickles in West, 'MDH-masala' in North etc., to name a few. And the food ingredients — nobody thought of. Only food ingredient that was there was tomato puree and paste in cans which in the recent times Godrej introduced in tetrapack.

Edible oil in tin pack and later on in polyester jar and tetrapack were the first major step in branding grocery items. "Postman"— peanut oil from Ahmed Mills was one of the oldest popular brands.

Today we get variety of edible oils and NDDB and Marico are the leading players in this sector. Amongst MNCs we have "Mazola" corn oil from CPC and "Sundrop" from ITC Agro — a sunflower oil which subsequently has been acquired by CONAGRA-the fastest growing food company in America. With CPC and CONAGRA being involved in edible oil market this sector is going to be soon dominated by MNCs.

The next major attempt to market a branded grocery item was done about a decade ago by Tatas by introducing the first refined iodised salt in poly bag by brand name "Tata Salt" and that was a success. The next big brand in this category is "Captain Cook" from DCW Chemicals who had plans in branded grocery items and thus the company introduced other items including wheat flour under the same brand name.

But that was shortlived. During 1997 DCW Chemicals suffered a major setback and the company decided to divest this food business. The acquirer is Corn Products Co. (India) Ltd. — a wholly owned subsidiary of CPC-International a nine billion US dollar American multinational in food and grocery business.

The company has recently changed its name to Best Foods Ltd. CPC was a sleepy company operating in India for over fifty years and their performance was lacklustre. With their limited range of Rex and Brown and Polson brands of convenience foods, the parent company, although doing well globally did not have any commitment in India.

In late seventies and early eighties when I was heading their project department many investment proposals were sent and notably among those were dextrose manufacturing project as CPC has a flavoured glucose brand in the portfolio and Knorr soup project. But project work used to be carried out by us endlessly but nothing got implemented.

In post liberalisation era, this apparently sleepy company seems to have become active. The third brand of salt is "Kissan Annapurna" which is making a sustained effort to get market share. This brand which is now in Hindustan Lever's fold through Brook Bond acquistion who in turn acquired Kissan in early nineties have reportedly spent Rs.18 crores in advertising and got 14% market share in a Rs.200 crore branded salt category in a span of one year.

Tata Salt still holds 26 percent market share and Captain Cook's share is about 20 percent and balance forty percent is still with the smaller players. With the acquisition of "Captain Cook" in branded salt market MNCs are the dominant players controlling sixty percent of the market.

The value addition in branded salt and pepper was really done successfully by a Delhi based local company Hi-Tech Foods belonging to Dharampal Satyapal of premium "Baba Zarda" (chewing tobacco) fame. Hi-Tech Food's `Catch' brand of salt and pepper in dispenser pack is a success story with upper middle class household and restaurant segments as the main customers. The sale of "Catch" is said to be around Rs.12 crores.

The first MNC to get into the branded spice business was Brook Bond who introduced the select premium priced spice range by name "Sona" in late eighties. The products were priced and packaged for higher income group.

They struggled for couple of years to establish "Sona" brand but failed and was forced to withdraw in later years. Apparently we don't see any major MNC in branded spice but world leader in exotic spices "Macomac" seems to have plan to enter Indian market. The first MNC to introduce Indian pickles was Nestle with their Maggie brand.

The products are still in the market. And as mentioned earlier, the first MNC to introduce traditional snacks (bujjias) is Pepsi under the umbrella of "Leher Namkins". It can be concluded that in traditional Indian foods MNCs cannot add much value through the involvement of their principals abroad. On the contrary, they will have to learn from the locals to derive advantage of their brand and resource muscle.

Hindustan Lever is expected to enter branded grocery items though Kissan route in a big way. The company does significant export of branded rice particularly in Middle East market and therefore expected to introduce the branded rice shortly in domestic market as well. Otherwise, in India so far branded packaged rice is limited to Basmati rice and there are many brands in this category and leading among them is "Kohinoor" — but all are from small manufacturers.

The grocery business is basically low margins and high volume business and thus it required multiple supply source to be strategically located to reduce the cost of freight and excellent distribution infrastructure and logistics management capability. Big houses like HLL, therefore are expected to do much better in this sector.

In the food ingredients business MNCs have not yet traded. The only big player in this area is Dabur with their range of pastes such as onion, garlic, ginger and also mix of these for particular application.

Current sales of these items which are sold under the umbrella brand "Hommade" is said to be Rs.5 crores and going by that standard it can be considered as a reasonable success. Very recently, "Cookme" curry paste has been launched in Calcutta. The market indicators point towards growth in the branded food ingredients business in coming years.

Pepsi is expected to get into this category and with their backward linkages with farmers whom they support with the high yielding variety of seeds for better crop yield, it stands to have an edge over others in this sector. One could have expected that Nestle would also jump into the fray in this sunrise food business — but their new Italian Managing Director Mr. Carlo Donati seems to have other priorities and they want to focus on major key brands.

Price and quality are expected to be the major determinant in the success of grocery items and basic foods. Consumers in India would not be willing to pay much as the price of convenience. Big global brands to MNCs' in that context are therefore not much of significance.

Imported Foods in Indian Market

Not too long ago for buying imported foods we had to go to select markets-such as Modern Bazar. Vasant Vihar in Delhi or Fort in Mumbai. In other places one has to literally search for such choice food items. In those days importing foods were not permitted.

And we were told that these foreign made chocolates and soup mixes and cheese find their way to Modern Bazar through diplomatic routes in the sense that persons in diplomatic services bring with them these foodstuff and sell them to local stores. But the fact of the matter is that these food items are brought to India through what is called "Suitcase trade" in the concerned circle. It will not take time for readers to understand what the "suitcase trade" is? It simply means to carry suitcase full of foodstuff by people frequently travelling overseas and selling and or delivering those to local traders and this happens in an organised manner. Those who can afford, have to pay a fancy price for those foreign foods. Although price was exhorbitant but the quality used to be distinctly better. Children love their Swiss chocolates, French Cheese and Italian pastas.

General Foods (now known as Kraft Foods International)-subsidiary of Phillips Morris has a synthetic orange beverage by brand-name "Tang" found to their dismay that so much of Tang is selling in India and Nepal now that they started thinking what would be the loss of sales to their South East Asian subsidiaries from where it finds its way to Indian market, once they start manufacturing it locally.

In the last couple of years these imported foods are available almost in all leading shops in most of the shopping centres in the city. To my surprise while taking a stroll in Greater Kailash shopping centre, I found couple of those shops have stacked Champaign, sparkling wine and imported mineral water outside the shop alongside soft drinks and on getting curious when I peeped inside I found everything is available. You name it — it is there. I bought some MARS chocolate and came out. The way things are moving, some of these foreign companies in course of time will do only trading in foods.

There are local agents who now represent many foreign companies and regularly import foods in turn to supply to local retailers and stores. In big stores like Nanz, there are regular requirement being catered to by their local agents.

In earlier years we used to see these phenomena in countries like say in Middle East. This trade in imported foods will be on the increase as the custom duty goes down. It will also fuel the growth of partial processing in India to take the advantage of price equation. For example, importing the product in bulk and carrying out re-packing operation here will be more economical than importing the finished product itself.

As we know that India has now signed the WTO (World Trade Organization) agreement, of course, being a signatory to this agreement is a lesser evil than not signing it at all. As per the terms of reference to this agreement, by year 2003, India has to put most of the items under Open General License (OGL) for import which means these items can be freely traded. We so far could announce around about hundred odd items under OGL and in that list shrimp, strawberry, rasberry and other fruits, have already featured. With the recent announcement of the Commerce Ministry, we have now freed import of 894 items including consumer goods and agricultural products by putting them under Open General Licence (OGL).

Another 414 items are removed from the "restricted" list allowing these to be imported against Special Import Licence (SIL) that can be procured in the market by importers at a small premium. Fruits like oranges and pomegranates, soft drink concentrates etc. are falling in SIL category whereas a range of marine food products, vegetables, assorted fruits and juices, spices, edible oils, jams, jellies etc. are brought under OGL.

The peak rate of the basic import tariff is now 40 percent which is applicable to most consumer products. On some items such as milk products India's WTO commitment bind it to a zero duty. But in most other agricultural products the WTO restriction is substantially higher ranging from 150 percent in the case of processed foods to 300 percent for edible oils. It is interesting to note that India, as a part of its globalisation policy is reducing the general tariff structure much faster than the asking rate as per WTO.

The outcome of the same is that in India we now get "KIWI" — an Australian fruit from select outlets if we want to. Shrimp is also now imported into India.

In a recent wealthy marwari marriage reception that I had attended, I found in the fruit counter all the fruits look different than what I am familiar about. For example, Guava and Apples were almost double the normal size that we get here. Plus there were unknown fruits. And I was told that all these fruits were imported from Bangkok.

I was so tempted that I tried all varieties and filled my stomach with all these tasty exotic fruits only. And later on started thinking that what will happen to our fruits if these fruits are available at a competitive price in India. Can our farmers and fruit merchants become globally competitive in terms of price and quality.

We should see the writing on the wall and work in that direction. The Agricultural Products Export Development Authority (APEDA) should quickly assess the scenario and rise to the occasion to take appropriate action. All of us are seeing the ongoing battle European banana growers are fighting in the forum of WTO against Americans refusal to permit entry of European Banana in that country.

If international fruit brands like "Chikita" and "Del Monte" start selling here, who is going to buy local fruits from our mandis and roadside vendors selling fruits on the top of hand-pulled trolleys?

Food Forum India: Redefining the Indian Food Business

The Food Forum India, the largest congregation of key, global and Indian, food retailers, manufacturers, organisations and minds in the food business will be held from May 6-7, 2008 at Renaissance, Mumbai. It will span the vast array of food business segments like food retail, food design, food processing, food logistics & support, food services, information & technology.

Conceptualised at the India Retail Forum (IRF), which saw the participation of leading brands, retailers and retail support organisations from across the globe, Food Forum India (FFI) was launched under the chairmanship of Subodh Kant Sahai, Union minister for food processing industries, with some 50 captains of the food world joining in to take up a series of support initiatives for the Indian food businesses.

The exhibition space covers all aspects of the food retail supply chain and has different exhibition zones such as Food Bites, Beverages, Food Support, Food Design, Food Space and Food Concepts.

F&G Retail Projections For 2010-11

- Reliance Fresh: To cover 70 cities, 784 urban towns and 6,000-odd rural mandis
- Future Group: 500 Spencer's Fresh, Daily, Super and Hyper
- Wadhawan Group: 1,500 Spinach/S-Mart/Sabka Bazaar outlets in 100 cities
- Subhiksha : 2,500 neighbourhood outlets across the nation

- Other major players firming up Pan-India plans include Bharti-Walmart, AV Birla Group (Trinethra/More), Heritage Foods (Fresh @, Hypercity, Dairy Farm, (Foodworld), Nilgiris, Radhakrishna Foodland, Influx, Landmark Group (Spar), Mother Dairy and Safal & region specific expansion by Jubilant Group (Jumbo Saver & Monday 2 Sunday), Hopcoms and Margin Free in South, Adanis and Arambagh Foodmart in East India among others. Most major retailers are expected to foray in to cash and carry model, to structure backend and take full advantage of large scale production, procurement, processing and also to put supply chain & logistics mechanisms in place.

Conferences : Several conferences will be held during the event.

Food Inspirations : The ability to do things differently has been one of the prerequisites to large scale success. Captains of successful food businesses will share their experiences and give insightful interpretations of the food industry.

Food Innovations : New ideas, products and concepts. Discussions on some of the most important innovations that have gained maximum popularity over the years. A look at possibilities for the near future.

Food For Thought : Presentations of leading global research and studies that are shaping the food industry today. Leading global food research and consulting firms and organisations will deliberate on key issues that guide the food business.

World Food Experience : The global market is today witnessing a mushrooming of numerous ideas and innovations that are radically altering the shape of businesses around the world. Focus will be on international ideas and insights, and include sunrise products from around the world and their implications for the Indian market.

Food Fight : Interaction between retailers and producers is of great importance to ensure the maintenance of an efficient chain that facilitates development of the industry as a whole. Food retailers and producers will debate on vital issues to determine the way ahead for the food industry in this session.

Investments In The Food Sector : Trade commissioners of key countries, governmental and regulatory bodies, international and domestic investors and corporates will debate upon investments in food processing, retailing, support and infrastructure.

Future Of Food In India 2020 : National and international vision leaders and key players of the industry will foresee the future of the business of food in India, and present some of the astounding possibilities of the future.

Awards : The Images Golden Spoon Awards (IGSA) to be announced at the event will honour enterprise, innovation and achievement in the business of food. Industry captains will celebrate excellence in the food business at a glamorous ceremony that will be the grand finale of Food Forum India.

The IGSA awards will be benchmark of excellence, that all food businesses will aspire for, thus playing a vital role in the in the progress of the industry. Food and grocery accounts for 63 per cent share in the Rs. 1,330,000 crore (US$ 300 million) Indian retail market. Only one per cent of this segment was organised in 2007. Food, grocery and beverages constitute about 43 per cent of India's private final consumption expenditure, and yet only 11 per cent of organised retail market in 2006.

Opportunity in Food and Grocery

The Urban Middle Class continues to grow by 20-25 million annually. The population is young, better exposed to modern lifestyles, and has a preference for packaged food and beverages. This growing Indian Middle Class, especially the employed youth in the ITES, BPO and other Service sectors, have higher propensity to consume, fuelled by easy availability of credit due to popularization of plastic cards.

More than 1,000 Food and Grocery outlets operational across tier-I and tier-II cities and Retail space of over 3 million sq. ft. have come up for F&G supermarkets, department stores and hypermarkets. 25 major F&G Retailers emerged in less than a year and Retail sales in this segment grew by nearly 150 per cent over 2006.

Opportunity Logistics

Improved logistics, warehousing, state-of-the-art technology and availability of quality processed and packaged food are factors that will boost the growth of organised F&G retail in India. Indian corporates like ITC, Godrej, DCM Shriram have initiated the process of direct supply linkages with the farm. Global major Metro AG was the first to introduce up-to-date know-how logistics and supply chain management. Corporate major Reliance is investing heavily on this front. Wal-Mart too plans back-end operations with Bharti.

Investments to the tune of US$ 23.5 billion in the food processing sector over the next three years is in the pipeline. This will undoubtedly create a huge demand for the food processing and packaging industry.. But keeping in view the vast farm sector and the surmounting demand for quality food products, there remains room for much more.

Opportunity in Food Processing

Nearly 80 per cent of agricultural produce in developed countries gets processed and packaged. It is just about 1.3 per cent in the case of India-a reason why less than one per cent of the F&G retail segment is organised. Even the existing organised F&G Retailers are finding it difficult to keep their stacks full & fresh.

This is just the right time for global firms to invest in the food processing sector, especially so when the government seems eager to boost rural/farm incomes. Marine Food Parks and Agricultural Special Economic Zones are already being worked out to ensure that farmers are not displaced. The opportunity is huge.. and early movers will reap the maximum advantage.

Opportunity to Boost the Incomes of Indian Farmers

Organised F&G retail requires direct sourcing of bulk produce and this ensures a boon for farmers. Initiatives thus far, are proving the point as traditional mandis loose relevance, middlemen get eased out, and as a result there is substantial increase in the income of farmers.

Adverse effects of fragmentation of land holdings can also be effectively countered. Initiative and Investment by Global companies can help make Cluster farming, Demand oriented farming and Raw material oriented farming the mainstay for Indian farmers.

India could well become the agri-business powerhouse in the export of fruits, vegetables, spices and food grains.

Opportunity for Food and Farm Supplies

More than 5,000 new outlets, 100 hypermarkets, 500 department stores and 2000 supermarkets are opening. Over 5,000 small and big existing outlets are to undergo a complete facelift. But the supplies are inadequate. Organised retail in India will need fresh and processed food and farm produce worth US$ 30 billion by the year 2010.

F&B (Servicing) Retail in India

Home-grown as well as international restaurant chains present in both high street locations and malls represent the organised food and beverages (F&B) services retail sector. This sector is in its growth phase and offers opportunities across a variety of retail formats.

The Indian F&B services sector is estimated to be worth Rs. 70,000 crore in 2007, at current prices, out of which about Rs. 5,750 crore is accounted for by the organised sector.

This is roughly 8.2 percent of the market being organised as compared to just 6.9 per cent of the market being organised in the previous year. This F&B services sector has increased its share from 4.7 per cent of the total of Rs. 1,200,000 crore retail market in 2006 to about 5.26 per cent of the Rs. 1,330,000 crore retail market in 2007, as per preliminary estimates by Images F&R research.

Report – India's Food Processing Industry

The Food Processing Industry sector in India is one of the largest in terms of production, consumption, export and growth prospects. The government has accorded it a high priority, with a number of fiscal reliefs and incentives, to encourage commercialisation and value addition to agricultural produce; for minimising pre/post harvest wastage, generating employment and export growth.

Important sub sectors in food processing industries are:-Fruit & Vegetable Processing, Fish-processing, Milk Processing, Meat & Poultry Processing, Packaged/Convenience Foods, Alcoholic beverages & Soft drinks and Grain Processing etc. As a result of several Policy Initiatives undertaken since liberalisation in August 1991, the industry has witnessed fast growth in most of the segments. As per a recent study on the food processing sector, the turnover of the total food market is approximately Rs.250,000 crores (US $ 69.4 billion) out of which value-added food products comprise Rs.80,000 crores (US $ 22.2 billion).

Since liberalisation in Aug 91 and up-till Feb 2000 proposals for projects of over Rs.53,800 crores (US.13.4 billion) have been proposed in various segments of the food and agro-processing industry. Besides this, Govt. has also approved proposals for joint ventures, foreign collaboration, industrial licenses and 100%export oriented units envisaging an investment of Rs.19,100 crores (US $ 4.80 billion) during the same period. Out of this, foreign investment is over Rs. 9100 crores (US $ 18.2 billion). Processed food exports were at over Rs.13,500 crores (US $ 3.2 billion) in 1998-99. Out of these exports, rice accounted for 46%, whereas marine products accounted for over 34%.

Primary food processing is a major industry with lakhs of rice-mills/hullers, flour mills, pulse mills and oil-seed mills. There are several thousands of bakeries, traditional food units and fruit/veg./ spice processing units in unorganised sector. In the organised sector, there are over 820 flour mills, 418 fish processing units, 5198 fruit/ veg processing units, 171 meat processing units.

India is the world's second largest producer of fruits & vegetables, but hardly 2% of the produce is processed. India is the land of spices producing all varieties worth over Rs. 3500 crores (US $ 900million) amounting to 25-30% of world production, which is processed for value-addition and export. It grows 22 million tonnes of oilseeds covering most of the varieties. Other important plantation products include tea, coffee, cocoa and cashew.

It has large marine product and processing potential with varied fish resources along the 8041 km. long coastline, 28000 km. of rivers and millions of hectares of reservoirs & brackish water. India's livestock population is largest in the world with 50% of world's buffaloes and 20% of cattles, but only about 1% of total meat production is converted to value added products. India is the largest milk producer in the world and about 15% of the total milk production is processed through the organised sector. Size of the semi-processed and ready to eat packaged food industry is over Rs. 4000 crores (US $ 1 billion) and is growing at over 20%.

With India's food industry witnessing explosive growth in line with the country's on-going economic development, there has been a marked move towards trying to get more out of the country's immense agricultural industry. Despite being the world's leading producer of cattle, milk, tea and pulses and the second largest producer of fruit, vegetables, rice, wheat and sugarcane worldwide, India accounts for just 1.5% of global processed food trade. The government estimates that the cost of boosting this figure to just 3% would be US$25bn, with India's primary focused agricultural and food supply chain having to dramatically reconsider standard industry practices in order to boost secondary output.

Fortunately, the government is not alone in attempting to fund this seismic shift. A vast population, steady urbanisation and GDP growth that is forecast at 7.9% annually to 2012, have helped ensure that multinational investment interest in the market is massive.

More encouraging than this is that multinationals alone are not being left to drive development; local corporations are also playing a huge role in the process. Mos notable has been the involvement of Indian companies in the food and beverage industry, whose primary business interest was formerly elsewhere. Over the past 12 months this has seen fast-moving consumer goods manufacturer Hindustan Lever – 52%-owned by Dutch giant Unilever – attempt to grow its food division, via the heavy marketing of popular brands, such as Knorr and Lipton, and reduce its dependence on its health and personal care products.

Perhaps even more significantly than this, burgeoning food sales have seen cigarette industry number one and two – ITC and Godfrey – launch new food products in a bid to establish themselves as key players within this industry.

Looking at the industry more broadly, almost all of India's leading conglomerates – Tata, Reliance, RPG, Aditya Birla etc. – are now participating in the industry to some degree, many having chosen the explosive grocery retail as a starting point. Likewise, former food category specialists, such as Amul Dairy, are now looking to become industry generalists in a bid to broaden their growth prospects (Amul has recently announced its interest in further non-dairy snack food launches).

Investment from all of these companies, as they battle to boost sales and build market share, should help to lift India's contribution to global processed food trade in the long term. In the short-to-medium term, the impact will be on internal processed food sales. Included in this report for the first time this quarter are our forecasts for the country's processed food industry. To 2012, we expect processed food output to increase by 44.2% to US$90.1bn, while packaged food sales in the country climb by 67.5% to US$21.7bn. On a per capita basis, it is forecast that a per capita packaged food spending increase of 56.5% to US$18.06 in 2012.

Food Processing – A New Era in India's Economy

The agriculture sector has come a long way since independence. With the advent of green revolution, India has transformed itself from a country of shortages to a land of surpluses. With the rapid growth of the economy, a shift is also being seen in the consumption pattern, from cereals to more varied and nutritious diet of fruit and vegetables, milk, fish, meat and poultry products.

This has resulted in the development of a sunrise industry namely the Food Processing Industries. The food processing sector in the country with its vast potential has emerged as one of the major driver of economic growth. It is encouraging to note that while the country's GDP growth rate had increase from 3.5 per cent in 2002-03 to 9 percent in 2006-07, the food processing sector has grown from 7 per cent to 13.1 per cent during the same period.

India is a country of over 1.10 billion consumers, there is a large untapped domestic market of 1,000 million consumers in the food processing sector and 200 million more consumers are expected to shift to processed food by 2010. It is the second largest producer of fruits and vegetables in the world.

There is a huge wastage of perishable food items in the country due to lack of proper food processing facilities and the level of processing is only about 2.2 per cent. However, India has tremendous potential to unleash large scale process based farm activities to exploit the emerging globalbusiness opportunities.

Incredible Opportunities

India's homogeneous market size endowed with growing incomes and changing life styles has created incredible market opportunities for food producers, machinery makers, food technology and service providers. Food processing industries has great export and employment potential. The policies are investor-friendly and more importantly technological and human resources are available aplenty in the country.

The competitive edge enjoyed in terms of raw material and labour offers lucrative opportunities. However poor perception of quality and the indifferent image of Indian products is preventing Indian food products to penetrate global markets in a big way. While developing countries like Thailand have exploited the global markets in a big was by fine-tuning quality management aspects of their food processing industry, India are yet to make a headway on this front.

Production of high quality processed foods meeting international quality standards & regulations may very well open new frontiers for Indian food products. This will not only create a dynamic and competitive domestic food processing industry but will also enable India to become a major player in the global food market. An attitudinal change towards quality is essential. Several thousand crore worth of farm produce is lost every year due to inefficient postharvest practices for storage and processing. On one hand is the growing demand for food products, which are difficult to meet due to limited resources and on the other, there exist abnormally high wastage in farm sector due to inefficient technology in storage, processing and handling.

It is, therefore, imperative to introduce state of the art technology in the food-processing sector to minimize post-harvest losses. It also calls for a concerted attention to a few selected food products where India has or can develop a competitive edge over other countries.

FDI

The sector has been attracting substantial FDI also and is among the top ten sectors getting FDI equity. FDI up to 100 per cent equity is permitted under the automatic route in food and infrastructure like food parks and cold chains. There are many areas for investment in this sector which include mega food parks, agri-infrastructure, supply chain aggregation, logistics and cold chain infrastructure, fruit and vegetable products, animal products, meat and dairy, fisheries and seafood cereals, consumer foods/ready to eat foods, wine and beer, machinery/packaging.

Productivity and Progress

It is essential to understand the dynamic relationship that exists between productivity and progress. The basic fact is that until both the farmers as well as the processors are convinced of benefits that accrue through productivity, the productivity campaigns will remain ineffective. So the main challenge is to introduce the concepts of productivity and make it work under a variety of constraints for the sustainable growth of the industry. In the process of globalization, the Indian food processing industry will be facing increased competition, particularly in domestic markets in addition to the uncertainties prevailing in the international markets.

It is in this context that emphasis must be given to improve productivity and quality. Undoubtedly, better performing firms will have a competitive edge over others. In order to maintain the tempo of productivity and quality, the National Productivity Awards have thus assumed much greater significance. Ministry of Food Processing Industries on its part, is leaving no stone unturned to achieve the multiple objectives of stepping up the growth, higher farmer income, reduction in wastage, providing nutritious and safe food and enhancing employment opportunities.

It has initiated measures to deal with the major constraints being faced by the industry such as affordability and cost of processed foods, linking of farmers and processors, supply chain and post harvest technology, infrastructure, finance, food safety, hygiene and taxes. With the active support and cooperation of all the stakeholders the ministry is confident of providing the necessary momentum for the rapid growth of the food processing sector and usher in a new era in the Indian economy.

Vision 2015

Given the strengths and opportunities of food processing sector, a Vision 2015 has been developed by the Ministry of Food Processing Industries, together with an appropriate strategy and implementable action plan so as to enhance farmer income, generate employment opportunities, provide choice to consumers at affordable price and contribute to overall national growth by increasing: the level of processing of perishables from 6% to 20%, value addition from 20% to 35% and share in global food trade from 1.5% to 3%.

The vision 2015 of the Government for the food processing sector aims at enhancing and stabilizing the income level of the farmers by assuring wider and better choice by enhancing dynamism, competitiveness, by ensuring safety and quality of food by introducing a transparent and scientific system of standards.

To achieve these aims a transparent and industry friendly regulatory regime is proposed to be established. Making the sector attractive for both domestic and foreign investors. Achieving integration of the food processing infrastructure from farm to market.

Having a transparent and industry friendly regulatory regime. Putting in place a transparent system of standards based on science. To achieve the stated Vision, Ministry of Food Processing Industries, has prepared Action Plan for energizing the food processing sector during 11th Plan.

Main Initiatives

Cold Chain: To address the situation and with a view to create a modern cold chain for preservation and value addition of perishables, during the 11th Plan, the Ministry is launching a revamped comprehensive Cold Chain Infrastructure Scheme for creating integrated cold chain infrastructure at different levels – farm level primary processing centre-cum-cold chain, collection/aggregation centres and Strategic Distribution Centres (SDC).

The SDCs will have integrated infrastructure facilities like material handling equipment, refrigeration, IQF/Blast freezing facility, Frozen/CA/MA Storage, Modern Packaging Facilities, ancillary equipment like X-ray, weigh bridge etc. The SDCs will be linked to retail supermarkets.

Mega Food Park : A new scheme of Mega Food Parks in the country is proposed which is envisaged to be a well defined agri/ horticultural processing zone containing state of the art processing facilities with support infrastructure and well established supply chain. The proposed scheme aims to provide a mechanism to bring together farmers, processors and retailers and link agricultural production to the market so as to ensure maximization of value addition, minimize wastages and improve farmers' income. The Mega Food Park is designed ultimately to link the farmers with the retail markets with minimizing of the intermediaries.

These food parks will function as sourcing hubs for the retail outlets. Abattoirs: Hygienic and scientific slaughtering as well as optimum utilization of byproducts are issues of grave concern of the Indian Meat Industry. It results in tremendous waste, contamination and avoidable cruelty to animals. Ministry is launching a comprehensive scheme for modernization of existing abattoirs/establishment of modern abattoirs at 100 locations across the country on a PPP mode.

Capacity Building: Ministry of Food Processing Industries has also taken up quality assurance, R & D, HACCP, Human Resource Development and Establishment of laboratories to support the Food Safety and Standards Act. Ministry of Food Processing Industries have taken many steps to give impetus to this sector which include virtual delicensing of the sector, inclusion in the priority sector for lending, allowing 100% FDI except in alcoholic beverages and retail, several duty and tax reliefs, financial assistance for infrastructure building, setting up of food processing units etc.

In case of export-oriented units, foreign investment is permitted even in case of items reserved for small scale sector. In addition, the export oriented units are given a number of incentives and concessions under the Export-Import Policy, such as, duty free import of capital goods, raw materials and intermediates, export income being exempt from Corporate Tax etc. FDI inflow in food processing is becoming stronger.

Recipe for Success

The global packaged food and nonalcoholic beverages industry, with more than 30 distinct product categories, is complex and highly fragmented—and worth about $2 trillion,or the rough equivalent of the GDP of the United Kingdom. The top 20 players account for less than one-quarter of the market, which is characterized by vast differences in local taste preferences, shelf-life considerations and food technologies.

All of these factors have combined to inhibit consolidation. However, deal activity has picked up in recent years as more companies have sought focus by trying to unravel diverse and undifferentiated portfolios. Yet even with increased consolidation, most companies are struggling to sustain margins.

Rates of growth and profitability vary significantly across the 30 product categories. Sales growth of bottled water far outstrips that of carbonates, for example. And sales of ready-made meals are almost double those of pasta. Such differences reflect rapidly evolving patterns of consumer demand—a reflection, in turn, of profound changes in demographics and lifestyle. More women in the workforce and increased urbanization have fuelled worldwide demand for convenience foods. At the same time, the rates of obesity and food-related diseases like diabetes have skyrocketed, powering a parallel desire for "healthy" food as well as a similarly health-related boom in organic foods, especially in developed countries.

Meanwhile, as consumers in many countries become wealthier, more informed and thus more confident, they have developed an interest in eating for pleasure and have cultivated a taste for exotic, ethnic foods. There are, to be sure, significant differences globally in these trends. Interest in healthy nutrition rises in line with disposable income, while the growth in convenience consumption is a function of both rising incomes and the increased sophistication of local retail networks. Emerging markets offer sizable growth opportunities, if only because of their relatively low per capita consumption rates.

The bulk of future growth, however, will come from North America, Western Europe and the Asia Pacific region, where the impact of the three new drivers of the demand—convenience, health and pleasure—is the strongest. High-performance businesses know this, and they have positioned themselves to take full advantage. Accenture defines a high-performance business as one that consistently outperforms its peers through economic and industry cycles as well as changes in leadership. Our extensive cross-industry research has identified three building blocks that underpin high performance in any industry—market focus and position; distinctive capabilities; and performance anatomy—and that all high-performance businesses have mastered. Here is what these building blocks look like in the food and nonalcoholic beverages industry.

Sidebar

About the Research

The peer group set for the study comprised 13 global, publicly traded companies that all boast worldwide sales of more than $5 billion annually. In addition, they all derive at least 75 percent of their sales from food and beverages, and 30 percent of these sales are outside their home region.

All provide branded packaged products (rather than food or beverage ingredients). Four companies emerged from our analysis as high-performance businesses. All distinguish themselves on measures of revenue growth, profitability, return on capital and working capital management. They also deliver superior economic profit, which has allowed them to produce consistently superior returns for their shareholders—more than double the industry average over the past five years.

More important, their returns on invested capital consistently outstrip those of their competitors, thanks to both higher margins and a higher percentage of sales per invested capital in each of the past three years. Indeed, high-performance businesses have grown nearly twice as fast as the industry average by focusing on high-growth categories and by implementing selective acquisitions to round out their portfolios and extend channel and geographic reach. However, only two of the four companies have consistently delivered superior returns to shareholders over both long and short time frames, as well as superior revenue growth and profitability and high future value expectations.

Case Study 1

PepsiCo: A discerning eye for acquisitions: Just 10 years ago, it looked as if PepsiCo had lost the so-called cola wars. Its profits lagged those of archrival Coca-Cola by almost 50 percent, and its market capitalization had slumped to less than half that of the Atlanta-based beverage giant.

Since 1996, PepsiCo has turned challenge into opportunity, reshaping its portfolio with an approach to acquisitions that has given the company strategic category domination—the essential element to succeeding at market focus and position, one of the three building blocks of high-performance business, in this industry. So much for history.

The key to PepsiCo's success is the selective nature of its M&A strategy. With the purchase of The Quaker Oats Co., South Beach Beverage Co. (SoBe) and Seagram Co.'s Tropicana Products division, the company has made strategic acquisitions in the water, sports drink and juice categories—all fast growing, thanks to the "healthy eating" trend that is one of the factors transforming the industry.

PepsiCo also has expanded (again selectively) into the packaged foods sector—more specifically, into snack foods, where its Frito-Lay division now enjoys a 60 percent share of the US market.

By acquiring top regional brands in attractive markets, PepsiCo is also huge—and growing—overseas. In February 2005, it took full control of a joint venture it had established with General Mills (said to be the largest of its kind in Europe) that makes and sells snack foods from Madrid to Moscow. Other recent acquisitions include Sara Lee Corp.'s nuts business in Belgium, the Netherlands and France, and a leading Polish snack foods company.

Case Study 2

Wrigley: Chewing up the competition: The Wm. Wrigley Jr. Co. has come a long way since 1892, when its founder first marketed a primitive form of chewing gum made from spruce gum and paraffin in Chicago. Today, the $4 billion company sells its popular brands in more than 180 countries. Indeed, with more than 60 percent of sales outside North America, Wrigley reigns supreme as the world's No. 1 maker of chewing and bubble gum.

These are hardly Wrigley's only products. Times have changed for a company that marketed just three chewing gums—Spearmint, Doublemint and Juicy Fruit—for more than 75 years. The growing popularity of sugar-free gum brands that promise to freshen breath, whiten teeth or fight tooth decay have made the gum market a target for candy companies like Wrigley's fast-growing rival, UK-based Cadbury Schweppes, which snapped up Pfizer's Adams confectionery unit, including the Dentyne and Bubblicious brands, in 2003.

Wrigley knows that insight-driven marketing—one of the important elements that distinguish high-performance businesses in this industry —is the key to sustaining its supremacy. So the company has been busily boosting product development.

Its innovation efforts often take the form of beefing up old favourites—pellet forms of Big Red and Juicy Fruit, for example. Wrigley invests heavily in its innovation capability and processes.

It spent $50 million to build a Global Innovation Centre—and it is reaping the rewards. Since 2001, the contribution from new products has averaged 15 percent to 20 percent of Wrigley's global sales—a significant increase over the single-digit figures of the 1990s.

Wrigley also has been on the selective acquisition trail—another key characteristic of high-performance businesses in this industry.

Historically, Wrigley focused exclusively on the gum subcategory and relied on organic means to grow the business. However, with increasing consolidation in the confectionery category and growing competitive pressures in the core gum market, Wrigley has responded with a string of acquisitions. In 2001, Wrigley bought Velamints and since then has been steadily adding to its range of non-gum mints, with products like Extra sugar-free mints.

In 2002, its $12.5 billion bid for the Hershey Company floundered when the sale was called off. In 2004, however, the purchase of the Spanish Joyco gum and confectionery business strengthened Wrigley's position in such key confectionery segments as bubble gum, candies and lollipops, and also boosted its presence in Europe and key emerging markets, including China and India.

In 2005, Wrigley resolutely entered the US sugar confectionery market with the acquisition of Kraft Foods' LifeSavers, Creme Savers and Altoids brands. Finally, in January 2007, the purchase of an 80 percent stake in Russian premium chocolatier A. Korkunov gave Wrigley an important foothold in a booming consumer marketplace.

Chapter 6

Food and Labour Cost Control

If you are running a restaurant or food service business, you understand that the most important costs under your control are food (including beverages) and labour costs — together, known in the industry as prime costs. Being able to compare these costs — in a percentage format — against typical scenarios of other restaurant businesses is helpful in the management of your business. Costs Vary Widely By Type of Restaurant

Both food and labour costs vary with the type of food service operation. As a rule, luxury restaurants will have higher food and labour cost percentages than casual dining or fast-food restaurants. The product sales mix, quality of food and service, pricing and hours of operation will impact your food and labour cost percentages.

Furthermore, state minimum wage differentials and differences in tip credit allowances (towards minimum wage) affect the labour cost percentage. The extent of beverages sales — as part of the food mix — has a considerable impact on total food cost percentages.

How Food and Labour Costs are Calculated

Food and labour costs are calculated as a percentage of the total volume of sales. If a restaurant does $20,000 per week and the total cost of food and beverages is $7,000 for that week, then the food cost is considered 35 percent. If, at the same restaurant, labour (including payroll taxes and benefits) equal $5,000 for the week, then the labour cost is 25 percent. Total prime costs are 60 percent in this example.

What Are the Ranges?

Certain fast food restaurants can achieve labour cost as low as 25 percent, while table service restaurants are more likely to see labour in the 30 percent to 35 percent range. Food costs (including beverages) for the restaurant industry run typically from the 25 percent to 38 percent range, depending upon the style of restaurant and the mix of sales.

Look at Prime Costs to Determine Success

In order to make money in the restaurant business, prime costs should generally be in the 60 percent to 65 percent range. How that breaks down between food and labour is less important than achieving a prime cost maximum that produces a satisfactory profit.

So if one of the prime costs is in the higher range, the other prime cost must be in the lower range to achieve profitability. Remember it is the combination of food and labour that creates the bottom line.

About Labour Costs

The majority of labour costs stem from paying employee wages and salaries. To obtain a more accurate picture of the restaurant's labour costs, divide the front-of-the-house labour with the back-of-the-house. Front-of-the-house labour includes wait staff, floor managers, hosts, bussers and bartenders. Back-of-the-house includes kitchen staff and dishwashers. As a small-business owner, also include the portion of all employee benefits that you pay, such as group health insurance, in your labour costs.

Labour Cost Percentages

Varying industry standards put labour costs between 30 to 35 percent of the restaurant's total sales. The standard for manager salaries should be around 10 percent of total labour costs, leaving 20 to 25 percent of labour costs for hourly employees. Labour costs tend to fluctuate depending on your restaurant's busy and slow seasons, so it's a good idea to average your weekly labour costs to obtain your monthly labour costs. You can then use your monthly labour cost averages to obtain your annual average labour costs.

Overstaffing

If you find your labour costs are higher than the industry standards because of overstaffing, examine your staff schedule and cut your staff down to a "skeleton crew" whenever possible. This applies to both front and back-of-the-house staff. If you have a kitchen manager or executive chef who schedules back-of-the-house staff, discuss your labour cost concerns and goals with her so she can schedule her staff accordingly.

Employee Productivity

Low employee productivity is also a factor in higher labour costs. To increase employee productivity and decrease the restaurant's labour costs, consider cross-training your front-of-the-house staff. For example, train some of your wait staff behind the bar. This way, on slow nights you won't have to staff a bartender because the server can take care of your guests' drink orders. For back-of-the-house employees, consider training them on additional food stations. Doing so will enable one person to work multiple stations on slow nights.

Controlling costs isn't so much about reducing them as it is about maximizing the value for your guests. Restaurateurs tend to look at theft, over portioning and shrinkage in terms of what it does to us as owners and managers. If we are to take a customer-centric approach to the problem, we begin with the question "What is the end result of these actions on my guests?"

Theft is an easy example. An employee that steals from you causes your Cost of Goods sold to increase which must be compensated for by raising prices. Raising prices in turn decreases the value of the customer's experience. The long term effect of this is that the customer count decreases which leads to lower sales.

A customer-centric look at over-portioning may at first look like a benefit to customers, but nothing is truly free, so if they aren't paying for it in the French fries on their plate, they'll pay for it with a beer price that could be priced 25 cents lower. While this may seem insignificant, the next point isn't so benign. Over-portioning is rarely consistent. This means that the customer that is delighted with the huge portion of French fries will be disappointed with a normal portion on their next visit. Major players in the fast food industry have shown how to handle this problem head on. McDonald's has changed their portions and packaging over the years to make it physically impossible to over-portion their fries.

This means that your fry box always seems to be spilling over with fries. It shows how seriously they take consistency in their business and gives us a glimpse at how they have become the world's number one operator.

Shrinkage or spoilage is the third issue that not only hits our bottom line but also our customers. Inefficient purchasing creates crisis situations (we need to get rid of this lettuce) and inferior product for your guests. Not ordering enough on the other hand creates shortages and also results in unhappy guest experiences.

Seven Strategies

So we've seen that these things hurt our guests as well as ourselves. What can we do to fix them?

Menu Development: Do you have a Products Standards Manual? If you do, do the portions listed reflect what's actually going on day-to-day. We've seen situations where the portion size listed in the manual was 40% less than what was actually going onto plates. Correcting this meant actually changing the size in the manual and adjusting pricing and purchasing decisions accordingly.

Portion Training: When a new team member comes on board, are they shown how to estimate portions? Do they check those estimates against a scale or other measure?

Prep Guides: How do you decide how much raw product to prep into your restaurant made sauces and other items? Key decisions like this should be thought out with all of the appropriate information at hand.

Purchasing: Reduce the number of vendors you deal with - How many suppliers do you work closely with? Have you spelled out expectations on produce, meats and other perishables?

Do you go back and forth between suppliers, focusing on price rather than quality? Instead, try focusing on one main supplier and establish strict expectations with them for quality.

Nearly all suppliers treat customers like that better than the "shoppers". By concentrating your volume, you'll gain leverage with the supplier as well.

Systemize Ordering: How do you decide how much product to order? Here's a hint, if you're sitting at a desk to do your order, it's probably wrong. Great operators walk through the restaurant's coolers, freezers, and dry storage while deciding how much to order.

In addition, they refer to the previous few weeks of usage to root out any trends or opportunities in their purchasing. As well, they see the condition of what's in stock and act accordingly.

Systemize Receiving: Do you have a process for receiving? There are four basic steps to receiving the order

- Check for completeness and accuracy
- Check for quality of product, especially produce and meats
- Contact your representative for any credits
- Date all products received.

Inventory Controls

Appropriate inventory controls are critical to closing the loop on controlling your costs. While they can be simple, the solution should be robust enough to alert you to trouble spots in your kitchen and on your menu.

Always keep in mind that your Cost of Goods Sold Percentage only tells you about a part of your business.

It won't tell you if your customers are unhappy with portion sizes or if your service levels are suffering. It's up to you as an excellent operator to balance all of the areas of your business.

Labour Cost

Operating a restaurant involves many expenses. Essentially, you incur labour costs when you have employees working for you.

Labour is an operational expense in just about any restaurant, predominantly including hourly wages for staff members.

Labour Cost Awareness

Labour costs are typically understood as a percentage of sales. To figure your labour cost, use this equation:

$$\textit{Payroll} \div \textit{Total Sales} = \textit{Labor Cost Percentage}$$

OR

$$\$7{,}000 \overline{)\$1{,}400}^{\;20\%}$$

Many restaurants hope to run a labour percentage below approximately 20 percent. When the costs begin to climb, anxiety levels rise as well.

Still, simply paying your employees less will not solve labour cost issues. The keys to controlling labour costs are improving workplace productivity and scheduling your employees wisely.

Control Your Labour Costs with Improved Worker Productivity

Increasing productivity improves your overall operation by building employee skills and confidence. Take time to provide your staff with sufficient training and communication.

Cross-Train Your Staff

Cross-training is beneficial to both the employee and the business, since the worker will have a wider range of skills and be able to help in multiple areas of the restaurant.

This allows the manager to schedule fewer workers while still being able to achieve the same production and service standards. Some suggestions for cross-training include:

- Train your prep cook to handle the grill
- Train your hostesses to work as back-up servers
- Train bussers to help run food to customers.

Conduct Frequent Staff Audits and Reviews

Another great way to help improve productivity is to perform regular audits. Take the time to watch and assess your employees' performances. If you find that a large portion of your employees' work days includes inordinately long breaks or downtime, it might be wise to revise your schedule. Conducting face-to-face reviews with each member of the staff will help communicate your thoughts and concerns.

Control Labour Cost With Precise Scheduling

Make sure you have constructed a budget to help keep track of your annual sales and expenses like labour. Through your budget, allow a percentage of your sales to cover labour expenses. Then, create a staffing schedule to reflect your budgeted allowance for labour expenses. The following tips elaborate these guidelines:

Break down your annual budget: Break down your annual budget into monthly budgets to help divide the money into weekly sections. This will give you a weekly budget, from which you can determine labour costs and make an appropriate staffing schedule.

Design a new weekly schedule for all employees: Relying on a fixed schedule week after week fails to acknowledge shifts in projected sales, changes in the weather or other factors that can affect your business. Adjust the number of staff scheduled each week to keep compliant with weekly budget constraints.

Monitor clock in/clock out times: After every shift, make sure that all employees have punched in and punched out exactly according to the schedule. Managers can usually use tools within the Point of Sale (POS) system to monitor and alter this information when necessary.

Discuss all schedule change requests in advance: Switching shifts can create problems when people start to work overtime, working more hours than the budget allows and potentially breaking a law, if the workers are youths. Be sure a manager is constantly aware of any proposed changes in the schedule.

Avoid Over-Staffing

It is often tempting to schedule more people than necessary in order to ensure that the business runs without any kinks. The reality is, however, that there will always be a few kinks in the restaurant business. Scheduling too many employees will increase your labour costs and reduce your overall profit, hurting your business overall. If you find that you have over-scheduled, you can send staff members home early. Train your people to work quickly, accurately and efficiently while also treating guests with respect and care. This allows you to operate at a high standard while still hitting your target labour percentage.

Whether opening a restaurant for the first time, or simply hiring due to employee turnover, be prepared to keep constant tabs on your staffing needs.

In the restaurant industry, a manager often needs to hire for the same position more than once. Learn to determine which positions you need to successfully run the business in order to find and keep the best staff members.

Know Your Staffing Needs

If you are opening a restaurant for the first time, you will need to determine your staffing needs with as much accuracy as possible. Even if you have run your business for a while and need to hire again, consider the following first:

- The type of restaurant you run (i.e.: fine dining, casual dining, fast food etc.)
- The type of food service you offer (i.e.: deli-style, buffet-style, fine-dining, etc.)
- The type of food on your menu—this determines what you need to cook and which kitchen stations you require
- The number of tables or seats available in your restaurant—this determines the number of servers you will need.

After you operate the restaurant for a while, you will begin to see the shifts and trends in daily and weekly business, and you will get a feel for which people—and how many people—you need for each part of the day.

Determine the Critical Positions for Your Operation

Know your concept before determining what position you will need to fill. For instance, a fine dining restaurant needs a chef and a sous chef, to start. However, a fast-food drive-thru will need a drive-thru order taker and fry-cooks.

Additionally, consider the positions that will have direct contact with customers or guests. Hiring respectful, friendly and engaging people will help define the face of your operation and deliver the memorable customer experience you want.

Determine How Many Employees to Hire

Before hiring, be sure to consider the number of employees you should bring on. It helps to start out with a few more employees than necessary in case some staff members do not work out. At the same time, be sure your annual budget can accommodate the number of people you are looking to hire. Essentially, consider the following guidelines:

Assess the number of tables, seats or service stations: The number of tables you expect to serve will help determine the number of servers you need on staff.

Consider the layout of the kitchen and different kitchen stations: The layout of your kitchen and the cooking stations affect the number of line cooks or kitchen staff you require.

Consider your busiest times first: Make sure your busiest times, typically lunch and dinner, are adequately staffed. For instance, you might need to staff more people from 11am-1pm than you would in early morning hours,

Stay within the means of your budget: Your budget will not only determine how much you can afford to pay your employees, but how many employees you can have working at a time to keep your labour cost down.

Use a Sales and Guest Projection Chart.

A staff chart is helpful when determining how many employees to schedule. The chart below shows the number of guests anticipated each hour, as well as the projected sales the restaurant achieve each hour.

This information determines the number of staff members needed per hour as well. You can organize your chart by number of expected tables if you do not know the expected hourly guest number. Be sure to consider your restaurant type, service style and layout when designing your chart, and keep operating efficiency and customer service as your priorities.

JOB POSITION	*# of Customer Per Hour*					
(# of each required)	*10*	*25*	*50*	*75*	*100*	*150*
Front of House						
Server	1	2	4	5	7	10
Busser	0	1	1	2	2	3
Bartender	1	1	1	1	2	2
Hostess	0	0	1	1	2	2
Back of House						
Line Cook	1	2	3	3	4	6
Dishwasher	1	1	1	1	1	2
Expeditor	0	0	0	1	1	1

Manage Your Labour Cost

When you begin to hire, or bring on more people, labour cost is a concern. Complete a comprehensive staffing schedule for all the staff you anticipate needing in your establishment. Cover all of your busy times and think about staffing according to the Sales and Guest Projection Chart.

For instance, you decide that you will need about four servers scheduled during the lunch hour. After you determine the average hourly wage for each position, you can simply multiply the average hourly wage by the hours required on your schedule. This is your labour cost. Compare this number with your operating budget, and make sure the labour cost does not displace profits. This will help you control your labour expenses.

Find the right people

Once you have considered your budget, the positions your restaurant needs, and how many people to hire, the real search begins. With a plan in place, you can focus on finding the right people for your operation and watch your operation become more successful before your eyes.

Controlling Food Costs

Controlling food costs starts with inventory organization and control. Over and over we have found that reducing inventory levels and organizing food storage areas will significantly reduce food costs. It is a simple axiom: When you have excess and disorganization you have waste.

The Best Guess Inventory/Order System

This system works best when your chef/kitchen manager has been able convince you that ordering is a complex procedure that you the manager/owner should not get involved in. Only someone with his/her many years of experience can understand the obscure secrets of food purchasing.

The system works like this:

- In the middle of the lunch hour rush a waitress runs out of crackers. (Side station stocking problem?)
- She rushes to the store room but does not see the crackers because two cases are hidden under other things. (Poor organization problem?)
- She then rushes to the chef, who is busy on the line, and tells him you are out of crackers.
- The chef was to busy to leave the line and check the storeroom.
- That afternoon whatever salesperson walks in first will get an order for crackers. (Forget price, we need them now!) And because we ran out the chef doubles the order to four cases.
- Now we have six cases of crackers that will sit in the storeroom for who knows how long. (Your money is sitting on a shelf earning nothing)
- Murphy's Law says it is only a matter of time until someone stacks a couple of cases on #10 cans on top of the crackers.
- Chances are good that before you can use up all those crackers, you will have one or more cases of cracker bits and crumbs.

There you are: Inadequate side station stocking and poor food storage organization caused a panic in the middle of the lunch rush resulting in a panic purchase of excess inventory, without checking prices, which resulted in overstocking and spoilage. The really scary part is that this is standard operating procedure for many restaurants.

Helpful Hint

In many instances we have eliminated 20% to 50% of the shelves and storage areas to encourage the kitchen to keep their inventory lean and mean. In some cases of extreme overstocking this alone has resulted in reducing food costs by more than 10%.

A rule of thumb we have found to work well is: Order to bring your inventory up to a two order period par level. That is: If you order an item daily, such as produce, order enough to have a two day supply including the inventory you have on hand. If you order dry stores weekly, subtract the inventory you have on hand from what you expect to use over the next two weeks and order the difference.

With this simple concept you should never run out of product, and never have more than a two order period inventory on hand. Check out the inventory/order system chapter for details.

Food Storage

While state Health Departments have gotten tougher in this area, many restaurants still loose product to freezer burn, contamination and spoilage because of improper storage procedures. Even worse are those restaurants who loose customers because they serve less than fresh meals.

Want to know what improper storage is costing you?

- On the sly, after all your employees have left, take an inventory of everything in your freezer and walk-in that is not properly wrapped, covered or date marked.
- In the next day or two take a second inventory to see how many of these items are still not properly stored. Check for rotation to see if the oldest items are used first.
- Now cost out the items that are still improperly stored. Multiply this by 365 days. Bet that number gets your attention.
- Give your kitchen crew a break by assuming that half of those items will be used up before they spoil. Chances are that even half of your annual spoilage cost will get you thinking what you could do with all that extra money.

Portion Control

Advance preparation can have a critical impact on food costs. Just as you would not stop to drain the swamp when the alligators are nipping at your butt, your cooks will not worry about portion control when deluged with orders during a rush.

Advance preparation and portioning not only saves food cost dollars, it also results in a more efficient kitchen, with a more consistent product and faster service to your customers.

Helpful Hint: A quick check of the previous weeks menu score provides a good guide as to how much to prep for a given day and meal period.

Purveyor Shopping

Regularly scheduled requests for bids/quotes from both your current suppliers and new suppliers can loop some more off your food costs. This is an area where you can delegate, up to a point.

- Have one or more of your employees make the calls, using the specifications from your inventory/order form. If a new supplier offers a good price, ask for samples.
- Do not let the employee change your specified supplier on the inventory/order form without your approval.
- Review the price offered and check the quality of the sample before changing purveyors. If you have any doubts, check their service and the consistency of their quality with other restaurants in your area.
- Have a receiving procedure that checks price, quality and quantity at your door. One way is to post your inventory/order form on a clipboard at the back door. The person receiving the delivery certifies that price and quantity of the delivery matches the inventory/order form as well as checking for quality. Exceptions can be noted on the back of the form, and signed by the delivery driver, with management follow-up.
- A double check is to have your bookkeeper check invoiced prices and quantities against quoted prices and amount ordered quantities on the inventory/order form

The Lowly Dishwasher

Being the lowest in the pecking order in a restaurant, the dishwasher's potential for profit maximizing is often overlooked. More than anyone else he/she can tell you what is coming back from your dining room. How much of your food costs are going into the trash because of excessive portions or just because customers do not like an entree, vegetable or condiment? Changing a garnish, eliminating a vegetable or reducing a portion may allow you to reduce the menu price, making you more price competitive with the same profit.

The Back Door

The back door is where a good portion your profits can go, if you let it.

- The most obvious step is to keep your back door locked. This will require you to install a door bell outside for deliveries. In addition local fire codes may require an emergency exit bar on the door. Be sure it includes an alarm that must be deactivated by management or designated kitchen personnel before the door can be opened.

- Keep the area outside your back door clean and tidy. As much as is practical you want to eliminate places where stolen food product can be hidden until the thief gets off shift.
- Check your garbage. It is a common trick to wrap a couple steaks up and hide them in the out going garbage. You may also realize considerable saving in lost linen and silverware costs by checking the garbage.
- If you are designing a new kitchen, locate your freezers and walk-ins, where you keep high cost items like meat and seafood, as far from the back door as possible. Make it harder to sneak things out.

How To Make It All Workable

Just like everything else in your restaurant, you are the key to making food cost control work. We favour the two person spot check and prosecute method.

The Two person check:

- Designate whoever has authority to open the back door to check all garbage before the dishwasher carries it out. If the inspector does his/her job theft will be reduced by requiring two person collusion.

The spot check:

- This is where management insures the inspector is doing his/her job. This is also where management gets their hands dirty and spot checks the garbage at unexpected times. This clearly establishes you are committed to controlling back door theft. It also validates that you do not ask your employees to do anything you will not do.

The prosecution:

- You may choose not to go through the hassle of calling the police and actually prosecuting a thief. What you can do is:
- If your cook/garbage inspector catches a dishwasher taking food out with the garbage, establish that the dishwasher is actually the thief and fire him/her. After a severe reprimand the first time, the same dishwasher caught taking food out again is reasonable proof of his/her guilt in our minds.

- If you wait outside and find stolen food in the garbage after your cook has inspected the garbage, fire the cook.
- It is very possible the cook who allows stolen food to go out the door is negligent rather than a thief. No one likes to root through garbage. By firing this cook you will insure that the next cook/inspector will really check the garbage.
- Help your cook/inspector. Provide heavy vinyl or rubber gloves and apron for use while inspecting the garbage.

A Couple Tricks To Reduce Breakage And Spoilage

- The very first step in breakfast preparation should be to set a supply of eggs out to temper, even before the grill is turned on. Egg yokes are less likely to break if the egg is cracked at room temperature.
- If you use frozen portion cut steaks, layer the steaks in a shallow hotel pan then cover them with clean cooking oil. With this simple procedure:
- The steak's cooler shelf life will be doubled. And, you will have better tasting steaks because the oil prevents the steaks flavourful juices from draining off.
- When broiled the oil seals in the steaks flavour, without leaving any after taste.
- With a longer shelf life you can keep more steaks defrosted, eliminating the need to defrost in a microwave, or attempt to broil a frozen steak if you sell more than expected.

Chapter 7

Food Technology

Food technology, or food tech for short is the application of food science to the selection, preservation, processing, packaging, distribution, and use of safe, nutritious, and wholesome food. Food scientists and food technologists study the physical, microbiological, and chemical make-up of food. Depending on their area of specialization, food scientists may develop ways to process, preserve, package, or store food, according to industry and government specifications and regulations. Consumers seldom think of the vast array of foods and the research and development that has resulted in the means to deliver tasty, nutritious, safe, and convenient foods., food technology is part of the curriculum and teaches, alongside *how* to cook, nutrition and the food manufacturing process.

Early History of Food Technology

Research in the field now known as food technology has been conducted for decades. Nicolas Appert's development in 1810 of the canning process was a decisive event. The process wasn't called canning then and Appert did not really know the principle on which his process worked, but canning has had a major impact on food preservation techniques. Louis Pasteur's research on the spoilage of wine and his description of how to avoid spoilage in 1864 was an early attempt to put food technology on a scientific basis. Besides research into wine spoilage, Pasteur did research on the production of alcohol, vinegar, wines and beer, and the souring of milk. He developed pasteurization—the process of heating milk and milk products to destroy food spoilage and disease-producing organisms. In his research into food technology, Pasteur became the pioneer into bacteriology and of modern preventive medicine. By 1940's to 1950's, the original four departments that had taught the subject under different names (including those at the University of Massachusetts and the University of California) had been retitled "food science", "food science and technology", or a similar variant.

Developments in Food Technology

Several companies in the food industry have played a role in the development of food technology. These developments have contributed greatly to the food supply. Some of these developments are:

- Instantized Milk Powder - D.D. Peebles (U.S. patent 2,835,586) developed the first instant milk powder, which has become the basis for a variety of new products that are rehydratable in cold water or milk. This process increases the surface area of the powdered product by partially rehydrating spray-dried milk powder.
- Freeze Drying - The first application of freeze drying was most likely in the pharmaceutical industry; however, a successful large-scale industrial application of the process was the development of continuous freeze drying of coffee.
- High-Temperature Short Time Processing - These processes for the most part are characterized by rapid heating and cooling, holding for a short time at a relatively high temperature and filling aseptically into sterile containers.
- Decaffeination of Coffee and Tea - Decaffeinated coffee and tea was first developed on a commercial basis in Europe around 1900. The process is described in U.S. patent 897,763. Green coffee beans are treated with steam or water to around 20% moisture.

  The added water and heat separate the caffeine from the bean to its surface. Solvents are then used to remove the caffeine from the beans. In the 1980s, new non-organic solvent techniques have been developed for the decaffeination of coffee and tea. Carbon dioxide under supercritical conditions is one of these new techniques. U.S. patent 4,820,537 was issued to General Foods Corp. for a CO_2 decaffeination process.
- Process optimization- Food Technology now allows production of foods to be more efficient, Oil saving technologies are now available on different forms. Production methods and methodology have also become increasingly sophisticated.

Technological Advancements in Food Products

Food Processing Industry Advances

America's current food-processing industry can be traced back to a tasking by Napoleon that led his quartermaster to engage the services of Nicolas Appert to produce stable rations for his troops to carry into battle. The canning industry that originated in 1809 has evolved during two centuries into a mature industry that still relies heavily on thermal processing to feed a growing population.

The Army used thermal processing to develop the C Ration, which evolved into the Meal, Combat Individual, that served U.S. troops from World War II through the Vietnam era. These rations relied on cylindrical metal cans.

In the late 1970s, the Army replaced metal cans with the more flexible and versatile polymeric retort pouch for improved thermally-processed products. The retort pouch was developed by the Natick Soldier Center, part of the U.S. Army Soldier Systems Center (Natick), in cooperation with industry and academia, and it is the central feature of the Meal, Ready-to-Eat (MRE). The 21st version of this ration, which will be procured in 2001, will have 24 different meals, each with a different entree in a retort pouch.

To increase the mobility of individual soldiers, lightweight, calorically-dense or dehydrated rations are required. The freeze-dehydration process pioneered by the Natick Soldier Center is capable of producing high-quality meals with long shelf life. This process produced the Long Range Patrol ration first used during the Vietnam conflict, and it is now used by industry for backpacker meals and survival rations, as well as for a variety of food ingredients. Current efforts are focused on cost-effective dehydration processes, such as osmotic dehydration and the development of ready-to-eat, partially dehydrated intermediate moisture items.

The military has continually exploited a variety of mechanical, chemical and electromagnetic energy sources and technologies for purposes of offense, defense or communication.

Numerous electromagnetic energy sources-initially developed for other purposes-can now be used for food and other materials processing. A variety of sources of energy across the electromagnetic spectrum may now be focused on food processing and other materials transformation mation processes. They are a dividend from generations of research aimed at much larger targets than the microorganisms that present threats to stability and safety of food materials. Currently, a considerable variety of thermal and nonthermal processes are being assessed to improve combat ration quality and variety. Among the thermal processes are high-temperature, short-time processes, including ohmic, microwave, radio frequency and induction heating.

The Natick Soldier Center and its precursor, the Quartermaster Food and Container Institute in Chicago, pioneered early research on nonthermal processes. The research explored food preservation potential of ionizing radiation (from both gamma and electron beam sources).

The Natick program was transferred to the USDA in 1980. However, data from the Natick project have been important in getting the approval of low-dose irradiation to improve the safety of fresh meat products and to extend the shelf life of fresh produce. Sterilized irradiated meat items prepared for NASA under Natick Soldier Center guidance are currently used in the Space Shuttle program.

Other pacing technologies envisioned for 21st century combat rations are two nonthermal processes, pulsed electric field (PEF) and high pressure (HP) preservation.

HP preservation can be applied to either liquid or solid foods in flexible containers. The improved food preservation technologies are being used to:

- Minimize processed-induced loss of color, flavour, texture, and nutrition.
- Retain the highest possible quality in stressful storage environments.
- Provide shelf-stable foods with fresh food attributes.
- Optimize ration variety, acceptance and consumption.
- Reduce the logistics burden and cost vs. conventionally-processed foods.
- Enhance the overall quality of life of the service member.

Because nonthermal processing is relatively new compared to conventional processing, two key issues are particularly critical for PEF and HP technologies. The first major issue is safety and process assurance; the second is sensory quality and consumer acceptance over the intended shelf life.

For military use that shelf life must be at least 3 years at 80 degrees Fahrenheit and 6 months at 100 degrees Fahrenheit. Several products made shelf-stable by nonthermal technologies were selected for microbiological and sensory assessment from the output of feasibility study contracts conducted 1992-1995. The contracts are as follows:

- Pulsed Electric Field (PEF)-FoodCo (now PurePulse Technologies), San Diego, Calif. Demonstration products were spaghetti sauce with meat, banana yogurt drink, milk and orange juice, a refrigerated product.
- Hyperbaric Pressure Preservation-Oregon State University and University of Delaware. Demonstration products were Spanish rice, spaghetti with meat sauce, fruit salad and yogurt with peaches.

After the products were shown to be microbiologically safe, sensory screenings were conducted against available, equivalent commercial products, military products or both using standard descriptive and/or consumer acceptance methods. As an example, batch-processed demonstration HP products submitted by Oregon State University were spaghetti with meat sauce and Spanish rice packed in 8-ounce polymeric bowls with aluminum tear-off closures treated in-package at 75,000 psi for 20 minutes. Despite the infancy of HPP processing, consumer panelists appearance, flavour and texture ratings of the spaghetti and rice items met or exceeded benchmark products both initially and after one year of storage at 80 F. A follow-on technology demonstration effort on HP processing was conducted under the leadership of Oregon State University.

One thrust was to optimize throughput and efficiency of a batch high-pressure processing system. Elmhurst Research, Inc., which turned to cannon technology for inspiration, developed a novel design for a batch processing system.

This effort has become a true "guns-to-plowshares" event in that a prototype system is now under development that uses recycled 6-inch cannon tubes as pressure vessels.

The Army's assessment of these novel technologies indicates that they do have considerable technology transfer potential to the food industry in view of the following:

- Consumer demand for minimally-processed foods with maximum nutrient retention.
- Continuing consumer demand for high quality and convenience along with assurance of safety.
- International competition in the industry.

As examples, ohmic (thermal) products are now commercial in Europe and Asia, and HP products are commercial in Japan. As noted earlier, the military is a potential beneficiary of technological advances in food processing to support integrated and focused logistics for battlefield operations and peacekeeping missions.

Finally, the role of the DoD Combat Ration Program is seen as serving as a catalyst for the food processing industry by securing considerable leveraging of resources through academic, industrial, government, and trade, technology or industry association partners in the United States and abroad. In this regard, the DoD has co-sponsored workshops of nonthermal researchers from industry, academia and government to jointly address the key technical barriers to commercialization of PEF and HP processing.

These workshops have led to the formation of a new Division of Nonthermal Processing for the Institute of Food Technologists.

The application of pulsed power technology to food preservation is being facilitated by a Fiscal Year 1999 Dual Use Science and Technology project.

The project for a Pulsed Electric Field Food Processing Consortium will be led by Ohio State University and industrial partners, including PurePulse, Inc.; TetraPak; Kraft; Nestles; and General Mills.

Industrial adoption of advanced food processing technologies will enhance the sustainment of our warfighters and will contribute to a more flexible and responsive logistics system.

Technological Advancements in Dairy Foods

Irradiation, biopreservatives and common sense ensure safety and extend shelflife. Capitol Hill food activists fall into two camps, those protesting advancements and applications in biotechnology and genetic engineering and those accusing processors of permitting unsafe food to enter the marketplace. Food safety is perhaps one of the greatest consumer concerns of the decade. According to a recent Gallup study commissioned by the American Meat Institute (AMI), of the 1,005 randomly selected adults interviewed nationwide, 79% were very or somewhat concerned about the safety of the food they purchased. These consumers are also extremely vulnerable to the activists on Capitol Hill.

Data from the Yankelovich Monitor indicates that from 1988 to 1993 consumer confidence in the media dropped from 53% to 23%, a 56% decrease. While their trust may be decreasing, the media exposure of food safety issues continues. Inevitably, consumers soak up prime time media, and respond accordingly. And as the media continues to report alleged health hazards in the American diet using descriptors such as "toxic," "carcinogenic" and "radioactive," consumers believe that the modern food supply is inherently dangerous due to the presence of synthetic chemicals and use of artificial ingredients and treatments.

The attached chart ranks food hazards, as calculated by FDA. The primary cause of illness from food is naturally occurring pathogens and their by-products. Pesticide residues and food additives are the least dangerous, but unfortunately, the media does not convey this message. For the most part, consumers do not understand that advancements in the areas activists protest are actually advancements toward a safer food supply.

Steps to Safety

"A Study of Consumers' Awareness, Knowledge and Attitudes Toward the Process of Irradiation," completed by AMI, revealed that the more consumers learn about food irradiation, the more they want it. "Spoilage organisms in dairy products are generally radiation-sensitive," explained Edward Remmers, v.p. of the American Council on Science and Health, at the recent Cultured Dairy Products Conference. "Unfortunately, radiation is not acceptable for dairy products promoting live and active cultures—they will be destroyed."

Remmers described the long-term recommendation made by Dr. E. Josephson, former head of U.S. Army's Food Irradiation Project and current adjunct professor of food science at the University of Rhode Island. "Explore irradiation of nonfat/lowfat cultured products at levels of 0.5, 1.0, 1.5 and 3.0 Kilograys to make a 2-3 logarithmic reduction in level of microorganisms in order to achieve a 2-3 fold extension of shelflife," says Remmers. "The greatest area of concern is the effect of radiation on radiation-sensitive sulphur-containing amino acids in milk proteins."

Mansel Griffiths, professor of food science, University of Guelph, Ontario, Canada, says, "Always equate shelflife with safety. And remember, there's no substitute for GMPs of HACCP." He explains that control of storage temperature and water activity are critical for the shelflife of yogurt. "Yogurt stored at 40 [degrees] F has a shelflife of about 35 days. Increase that to 50 [degrees] F, and shelflife reduces to 19 days. And the lower the water activity, the longer the shelflife."

Overcoming High Moisture

Researchers at the University of Wisconsin-Madison Food Research Institute continue to assess the quality and safety of dairy foods. During a recent meeting with Eric Johnson, associate professor, and John Luchansky, assistant professor, both from the department of food microbiology and toxicology; and Mark Johnson, senior scientist at the Center for Dairy Research, they agreed that both starter and nonstarter bacteria flourish in higher moisture, lower fat cheeses.

Growth of undesirable microbes in lower fat cheese results in faster protein breakdown (softer body), gassing and/or off-flavour development. They emphasize concentrating on the following variables in cheese manufacturing, especially in reduced fat versions, to ensure safety and shelflife.

- Pasteurization of milk is an absolute must for reduced-fat and fat-free cheeses.
- Good Manufacturing Practices will keep coliforms out.
- Biofilm development can be prevented by properly cleaning and sanitizing vats, eliminating the transfer and growth of undesirable non-pathogenic bacteria.
- Salt to water ratio must be properly maintained in order to inhibit the growth of nonstarter bacteria. Unfortunately, an increase in salt also inhibits starter bacteria, and slows protein breakdown and body development, resulting in a salty bland cheese. Salt substitutes may prove beneficial.
- Strict control of bacteria in the manufacturing and storage areas can reduce the numbers of non-starter bacteria. Cold storage temperatures can inhibit growth.
- Efficient distribution should be incorporated into the market strategy for higher moisture cheeses due to flavour and texture breakdown.

Natural Preservatives

Luchansky's research group is currently conducting extensive research on the effects of biopreservatives in natural cheese. While this concept of utilizing lactic culture and their metabolites as a preservative has been around for years, it has increased in the last five years due to pressures from activists against other forms of safety assurance.

The bacteriocins (natural preservatives derived from bacteria) evaluated most extensively in dairy foods are nisin (produced by

Lactococcus lactis) and pediocin (produced by Pediococcus acidilactici). "Our group has incorporated bacteriocins at various stages during cheese manufacture to control both pathogenic and spoilage bacteria," says Luchansky. "For example, in the presence of increasing levels of pediocin, numbers of Listeria monocytogenes decreased by 1.5 to 4.0 units compared to control treatments." This experiment involved treated Cheddar cheese samples held at 39.2 [degrees] F for 35 days.

Luchansky suggests selecting or constructing bacteriocin-producing cultures for use as starter or starter adjuncts in the cheese make process. Or, commercially available fermentates of lactic acid bacteria can be added to the milk or to the curd with the salt. Additional research includes bacteriocin applications by means of packaging material and surface application.

Ranking of Food Hazards (Most to Least Dangerous)

1. Microbial contamination
2. Naturally occurring toxicants
3. Environmental contaminants (metals)
4. Nutritional problems (malnutrition, undernutrition)
5. Pesticide residues
6. Food additives.

During the Middle Ages, farmers in North Africa, the Near East, and Europe began making use of agricultural technologies including irrigation systems based on hydraulic and hydrostatic principles, machines such as norias, water-raising machines, dams, and reservoirs. This combined with the invention of a three-field system of crop rotation and the moldboard plow greatly improved agricultural efficiency.

In the European medieval period, agriculture was considered part of the set of *seven mechanical arts*.

Modern Era

After 1492, a global exchange of previously local crops and livestock breeds occurred. Key crops involved in this exchange included the tomato, maize, potato, manioc, cocoa bean and tobacco going from the New World to the Old, and several varieties of wheat, spices, coffee, and sugar cane going from the Old World to the New.

The most important animal exportation from the Old World to the New were those of the horse and dog (dogs were already present in the pre-Columbian Americas but not in the numbers and breeds suited to farm work). Although not usually food animals, the horse (including donkeys and ponies) and dog quickly filled essential production roles on western-hemisphere farms.

The potato became an important staple crop in northern Europe. Since being introduced by Portuguese in the 16th century, maize and manioc have replaced traditional African crops as the continent's most important staple food crops. By the early 19th century, agricultural techniques, implements, seed stocks and cultivated plants selected and given a unique name because of its decorative or useful characteristics had so improved that yield per land unit was many times that seen in the Middle Ages. Although there is a vast and interesting history of crop cultivation before the dawn of the 20th century, there is little question that the work of Charles Darwin and Gregor Mendel created the scientific foundation for plant breeding that led to its explosive impact over the past 150 years.

With the rapid rise of mechanization in the late 19th century and the 20th century, particularly in the form of the tractor, farming tasks could be done with a speed and on a scale previously impossible. These advances have led to efficiencies enabling certain modern farms in the United States, Argentina, Israel, Germany, and a few other nations to output volumes of high-quality produce per land unit at what may be the practical limit.

The Haber-Bosch method for synthesizing ammonium nitrate represented a major breakthrough and allowed crop yields to overcome previous constraints. In the past century agriculture has been characterized by enhanced productivity, the substitution of labour for synthetic fertilizers and pesticides, water pollution, and farm subsidies. In recent years there has been a backlash against the external environmental effects of conventional agriculture, resulting in the organic movement.

The cereals rice, corn, and wheat provide 60% of human food supply. Between 1700 and 1980, "the total area of cultivated land worldwide increased 466%" and yields increased dramatically, particularly because of selectively bred high-yielding varieties, fertilizers, pesticides, irrigation, and machinery. For example, irrigation increased corn yields in eastern Colorado by 400 to 500% from 1940 to 1997. However, concerns have been raised over the sustainability of intensive agriculture. Intensive agriculture has become associated with decreased soil quality in India and Asia, and there has been increased concern over the effects of fertilizers and pesticides on the environment, particularly as population increases and food demand expands. The monocultures typically used in intensive agriculture increase the number of pests, which are controlled through pesticides. Integrated pest management (IPM), which "has been promoted for decades and has had some notable successes" has not significantly affected the use of pesticides because policies encourage the use of pesticides and IPM is knowledge-intensive.

Although the "Green Revolution" significantly increased rice yields in Asia, yield increases have not occurred in the past 15–20 years. The genetic "yield potential" has increased for wheat, but the yield potential for rice has not increased since 1966, and the yield potential for maize has "barely increased in 35 years". It takes a decade or two for herbicide-resistant weeds to emerge, and insects become resistant to insecticides within about a decade. Crop rotation helps to prevent resistances.

Agricultural exploration expeditions, since the late 19th century, have been mounted to find new species and new agricultural practices in different areas of the world. Two early examples of expeditions include Frank N. Meyer's fruit- and nut-collecting trip to China and Japan from 1916-1918 and the Dorsett-Morse Oriental Agricultural Exploration Expedition to China, Japan, and Korea from 1929-1931 to collect soybean germplasm to support the rise in soybean agriculture in the United States.

In 2009, the agricultural output of China was the largest in the world, followed by the European Union, India and the United States, according to the International Monetary Fund. Economists measure the total factor productivity of agriculture and by this measure agriculture in the United States is roughly 2.6 times more productive than it was in 1948.

Six countries - the US, Canada, France, Australia, Argentina and Thailand - supply 90% of grain exports. The United States controls almost half of world grain exports. Water deficits, which are already spurring heavy grain imports in numerous middle-sized countries, including Algeria, Iran, Egypt, and Mexico, may soon do the same in larger countries, such as China or India.

Crop Production Systems

Cropping systems vary among farms depending on the available resources and constraints; geography and climate of the farm; government policy; economic, social and political pressures; and the philosophy and culture of the farmer. Shifting cultivation (or slash and burn) is a system in which forests are burnt, releasing nutrients to support cultivation of annual and then perennial crops for a period of several years. Then the plot is left fallow to regrow forest, and the farmer moves to a new plot, returning after many more years (10-20). This fallow period is shortened if population density grows, requiring the input of nutrients (fertilizer or manure) and some manual pest control. Annual cultivation is the next phase of intensity in which there is no fallow period. This requires even greater nutrient and pest control inputs. Further industrialization lead to the use of monocultures, when one cultivar is planted on a large acreage. Because of the low biodiversity, nutrient use is uniform and pests tend to build up, necessitating the greater use of pesticides and fertilizers. Multiple cropping, in which several crops are grown sequentially in one year, and intercropping, when several crops are grown at the same time are other kinds of annual cropping systems known as polycultures.

In tropical environments, all of these cropping systems are practiced. In subtropical and arid environments, the timing and extent of agriculture may be limited by rainfall, either not allowing multiple annual crops in a year, or requiring irrigation. In all of these environments perennial crops are grown (coffee, chocolate) and systems are practiced such as agroforestry. In temperate environments, where ecosystems were predominantly grassland or prairie, highly productive annual cropping is the dominant farming system.

The last century has seen the intensification, concentration and specialization of agriculture, relying upon new technologies of agricultural chemicals (fertilizers and pesticides), mechanization, and plant breeding (hybrids and GMO's). In the past few decades, a move towards sustainability in agriculture has also developed, integrating ideas of socio-economic justice and conservation of resources and the environment within a farming system. This has led to the development of many responses to the conventional agriculture approach, including organic agriculture, urban agriculture, community supported agriculture, ecological or biological agriculture, integrated farming and holistic management, as well as an increased trend towards agricultural diversification.

Crop Statistics

Important categories of crops include grains and pseudograins, pulses (legumes), forage, and fruits and vegetables. Specific crops are cultivated in distinct growing regions throughout the world. In millions of metric tons, based on FAO estimate.

Livestock Production Systems

Animals, including horses, mules, oxen, camels, llamas, alpacas, and dogs, are often used to help cultivate fields, harvest crops, wrangle other animals, and transport farm products to buyers. Animal husbandry not only refers to the breeding and raising of animals for meat or to harvest animal products (like milk, eggs, or wool) on a continual basis, but also to the breeding and care of species for work and companionship. Livestock production systems can be defined based on feed source, as grassland - based, mixed, and landless. Grassland based livestock production relies upon plant material such as shrubland, rangeland, and pastures for feeding ruminant animals. Outside nutrient inputs may be used, however manure is returned directly to the grassland as a major nutrient source.

This system is particularly important in areas where crop production is not feasible because of climate or soil, representing 30-40 million pastoralists. Mixed production systems use grassland, fodder crops and grain feed crops as feed for ruminant and monogastic (one stomach; mainly chickens and pigs) livestock. Manure is typically recycled in mixed systems as a fertilizer for crops. Approximately 68% of all agricultural land is permanent pastures used in the production of livestock.

Landless systems rely upon feed from outside the farm, representing the de-linking of crop and livestock production found more prevalently in OECD member countries. In the U.S., 70% of the grain grown is fed to animals on feedlots. Synthetic fertilizers are more heavily relied upon for crop production and manure utilization becomes a challenge as well as a source for pollution.

Production Practices

Tillage is the practice of plowing soil to prepare for planting or for nutrient incorporation or for pest control. Tillage varies in intensity from conventional to no-till. It may improve productivity by warming the soil, incorporating fertilizer and controlling weeds, but also renders soil more prone to erosion, triggers the decomposition of organic matter releasing CO_2, and reduces the abundance and diversity of soil organisms.

Pest control includes the management of weeds, insects/mites, and diseases. Chemical (pesticides), biological (biocontrol), mechanical (tillage), and cultural practices are used. Cultural practices include crop rotation, culling, cover crops, intercropping, composting, avoidance, and resistance. Integrated pest management attempts to use all of these methods to keep pest populations below the number which would cause economic loss, and recommends pesticides as a last resort.

Nutrient management includes both the source of nutrient inputs for crop and livestock production, and the method of utilization of manure produced by livestock. Nutrient inputs can be chemical inorganic fertilizers, manure, green manure, compost and mined minerals.

Crop nutrient use may also be managed using cultural techniques such as crop rotation or a fallow period. Manure is used either by holding livestock where the feed crop is growing, such as in managed intensive rotational grazing, or by spreading either dry or liquid formulations of manure on cropland or pastures.

Water management is where rainfall is insufficient or variable, which occurs to some degree in most regions of the world. Some farmers use irrigation to supplement rainfall. In other areas such as the Great Plains in the U.S. and Canada, farmers use a fallow year to conserve soil moisture to use for growing a crop in the following year. Agriculture represents 70% of freshwater use worldwide.

Processing, Distribution, and Marketing

In the United States, food costs attributed to processing, distribution, and marketing have risen while the costs attributed to farming have declined. This is related to the greater efficiency of farming, combined with the increased level of value addition (e.g. more highly processed products) provided by the supply chain. From 1960 to 1980 the farm share was around 40%, but by 1990 it had declined to 30% and by 1998, 22.2%. Market concentration has increased in the sector as well, with the top 20 food manufacturers accounting for half the food-processing value in 1995, over double that produced in 1954. As of 2000 the top six US supermarket groups had 50% of sales compared to 32% in 1992. Although the total effect of the increased market concentration is likely increased efficiency, the changes redistribute economic surplus from producers (farmers) and consumers, and may have negative implications for rural communities.

Crop Alteration and Biotechnology

Crop alteration has been practiced by humankind for thousands of years, since the beginning of civilization. Altering crops through breeding practices changes the genetic make-up of a plant to develop crops with more beneficial characteristics for humans, for example, larger fruits or seeds, drought-tolerance, or resistance to pests. Significant advances in plant breeding ensued after the work of geneticist Gregor Mendel.

His work on dominant and recessive alleles gave plant breeders a better understanding of genetics and brought great insights to the techniques utilized by plant breeders. Crop breeding includes techniques such as plant selection with desirable traits, self-pollination and cross-pollination, and molecular techniques that genetically modify the organism.

Domestication of plants has, over the centuries increased yield, improved disease resistance and drought tolerance, eased harvest and improved the taste and nutritional value of crop plants. Careful selection and breeding have had enormous effects on the characteristics of crop plants.

Plant selection and breeding in the 1920s and 1930s improved pasture (grasses and clover) in New Zealand. Extensive X-ray an ultraviolet induced mutagenesis efforts (i.e. primitive genetic engineering) during the 1950s produced the modern commercial varieties of grains such as wheat, corn (maize) and barley. The Green Revolution popularized the use of conventional hybridization to increase yield many folds by creating "high-yielding varieties".

For example, average yields of corn (maize) in the USA have increased from around 2.5 tons per hectare (t/ha) (40 bushels per acre) in 1900 to about 9.4 t/ha (150 bushels per acre) in 2001. Similarly, worldwide average wheat yields have increased from less than 1 t/ha in 1900 to more than 2.5 t/ha in 1990. South American average wheat yields are around 2 t/ha, African under 1 t/ha, Egypt and Arabia up to 3.5 to 4 t/ha with irrigation.

In contrast, the average wheat yield in countries such as France is over 8 t/ha. Variations in yields are due mainly to variation in climate, genetics, and the level of intensive farming techniques (use of fertilizers, chemical pest control, growth control to avoid lodging).

Genetic Engineering

Genetically Modified Organisms (GMO) are organisms whose genetic material has been altered by genetic engineering techniques generally known as recombinant DNA technology. Genetic engineering has expanded the genes available to breeders to utilize in creating desired germlines for new crops. After mechanical tomato-harvesters were developed in the early 1960s, agricultural scientists genetically modified tomatoes to be more resistant to mechanical handling.

More recently, genetic engineering is being employed in various parts of the world, to create crops with other beneficial traits. New research on woodland strawberry genome was found to be short and easy to manipulate. Researchers now have tools to improve strawberry flavours and aromas of cultivated strawberries as stated in a publication by Nature Genetics.

Herbicide-tolerant GMO Crops

Roundup Ready seed has a herbicide resistant gene implanted into its genome that allows the plants to tolerate exposure to glyphosate. *Roundup* is a trade name for a glyphosate-based product, which is a systemic, nonselective herbicide used to kill weeds. *Roundup Ready* seeds allow the farmer to grow a crop that can be sprayed with glyphosate to control weeds without harming the resistant crop. Herbicide-tolerant crops are used by farmers worldwide.

Today, 92% of soybean acreage in the US is planted with genetically modified herbicide-tolerant plants. With the increasing use of herbicide-tolerant crops, comes an increase in the use of glyphosate-based herbicide sprays.

In some areas glyphosate resistant weeds have developed, causing farmers to switch to other herbicides. Some studies also link widespread glyphosate usage to iron deficiencies in some crops, which is both a crop production and a nutritional quality concern, with potential economic and health implications.

Insect-resistant GMO Crops

Other GMO crops used by growers include insect-resistant crops, which have a gene from the soil bacterium *Bacillus thuringiensis* (Bt), which produces a toxin specific to insects. These crops protect plants from damage by insects; one such crop is Starlink. Another is cotton, which accounts for 63% of US cotton acreage.

Some believe that similar or better pest-resistance traits can be acquired through traditional breeding practices, and resistance to various pests can be gained through hybridization or cross-pollination with wild species. In some cases, wild species are the primary source of resistance traits; some tomato cultivars that have gained resistance to at least nineteen diseases did so through crossing with wild populations of tomatoes.

Costs and Benefits of GMOs

Genetic engineers may someday develop transgenic plants which would allow for irrigation, drainage, conservation, sanitary engineering, and maintaining or increasing yields while requiring fewer fossil fuel derived inputs than conventional crops.

Such developments would be particularly important in areas which are normally arid and rely upon constant irrigation, and on large scale farms. However, genetic engineering of plants has proven to be controversial.

Many issues surrounding food security and environmental impacts have risen regarding GMO practices. For example, GMOs are questioned by some ecologists and economists concerned with GMO practices such as terminator seeds, which is a genetic modification that creates sterile seeds. Terminator seeds are currently under strong international opposition and face continual efforts of global bans.

Another controversial issue is the patent protection given to companies that develop new types of seed using genetic engineering. Since companies have intellectual ownership of their seeds, they have the power to dictate terms and conditions of their patented product. Currently, ten seed companies control over two-thirds of the global seed sales.

Vandana Shiva argues that these companies are guilty of biopiracy by patenting life and exploiting organisms for profit Farmers using patented seed are restricted from saving seed for subsequent plantings, which forces farmers to buy new seed every year. Since seed saving is a traditional practice for many farmers in both developing and developed countries, GMO seeds legally bind farmers to change their seed saving practices to buying new seed every year.

Locally adapted seeds are an essential heritage that has the potential to be lost with current hybridized crops and GMOs. Locally adapted seeds, also called land races or crop eco-types, are important because they have adapted over time to the specific microclimates, soils, other environmental conditions, field designs, and ethnic preference indigenous to the exact area of cultivation. Introducing GMOs and hybridized commercial seed to an area brings the risk of cross-pollination with local land races.

Therefore, GMOs pose a threat to the sustainability of land races and the ethnic heritage of cultures. Once seed contains transgenic material, it becomes subject to the conditions of the seed company that owns the patent of the transgenic material.

Modern Agriculture

Modern agriculture is a term used to describe the wide majority of production practices employed by America's farmers. The term depicts the push for innovation, stewardship and advancements continually made by growers to sustainably produce higher-quality products with a reduced environmental impact.

Intensive scientific research and robust investment in modern agriculture during the past 50 years has helped farmers double food production while essentially freezing the footprint of total cultivated farmland.

Safety

The agriculture industry works with government agencies and other organizations to ensure that farmers have access to the technologies required to support modern agriculture practices. Farmers are supported by education and certification programs that ensure they apply agricultural practices with care and only when required.

Sustainability

Technological advancements help provide farmers with tools and resources to help reduce their environmental footprint and to make farming more sustainable.

New technologies have given rise to innovations like conservation tillage, a farming process which helps prevent land loss to erosion, water pollution and enhances carbon sequestration.

The World Bank, the Bill & Melinda Gates Foundation and others have noted that integrated crop management is based on agro-ecological principles and can increase yields while reducing environmental damage.

Affordability

The goal of modern agriculture practices is to help farmers provide an affordable supply of food to meet the demands of a growing population. With modern agriculture, more crops can be grown on less land allowing farmers to provide an increased supply of food at an affordable price.

Food Safety, Labelling and Regulation

Food security issues also coincide with food safety and food labelling concerns. Currently a global treaty, the BioSafety Protocol, regulates the trade of GMOs. The EU currently requires all GMO foods to be labelled, whereas the US does not require transparent labelling of GMO foods. Since there are still questions regarding the safety and risks associated with GMO foods, some believe the public should have the freedom to choose and know what they are eating and require all GMO products to be labelled.

The Food and Agriculture Organization of the United Nations (FAO) leads international efforts to defeat hunger and provides a neutral forum where nations meet as equals to negotiate agreements and debate food policy and the regulation of agriculture. According to Dr. Samuel Jutzi, director of FAO's animal production and health division, lobbying by "powerful" big food corporations has stopped reforms that would improve human health and the environment.

The "real, true issues are not being addressed by the political process because of the influence of lobbyists, of the true powerful entities," he said, speaking at the Compassion in World Farming annual forum.

For example, recent proposals for a voluntary code of conduct for the livestock industry that would have provided incentives for improving standards for health, and environmental regulations, such as the number of animals an area of land can support without long-term damage, were successfully defeated due to large food company pressure.

Environmental Impact

Agriculture imposes external costs upon society through pesticides, nutrient runoff, excessive water usage, and assorted other problems. A 2000 assessment of agriculture in the UK determined total external costs for 1996 of £2,343 million, or £208 per hectare.

A 2005 analysis of these costs in the USA concluded that cropland imposes approximately $5 to 16 billion ($30 to $96 per hectare), while livestock production imposes $714 million. Both studies concluded that more should be done to internalize external costs, and neither included subsidies in their analysis, but noted that subsidies also influence the cost of agriculture to society. Both focused on purely fiscal impacts.

The 2000 review included reported pesticide poisonings but did not include speculative chronic effects of pesticides, and the 2004 review relied on a 1992 estimate of the total impact of pesticides.

Agriculture accounts for 70 per cent of withdrawals of freshwater resources. However, increasing pressure being placed on water resources by industry, cities and the involving biofuels industry means that water scarcity is increasing and agriculture is facing the challenge of producing more food for the world's growing population with fewer water resources.

Scientists are also realising that water resources need to be allocated to maintain natural environmental services, such as protecting towns from flooding, cleaning ecosystems and supporting fish stocks. In the book *Out of Water: From abundance to scarcity and how to solve the world's water problems*, authors Colin Chartres and Samyukta Varma of the International Water Management Institute lay down a six-point plan of action for addressing the global challenge of producing sufficient food for the world with dwindling water resources.

One of the actions they say is required is to ensure all water systems, such as lakes and rivers, have water allocated to environmental flow. A key player who is credited to saving billions of lives because of his revolutionary work in developing new agricultural techniques is Norman Borlaug. His transformative work brought high-yield crop varieties to developing countries and earned him an unofficial title as the father of the Green Revolution.

Livestock Issues

A senior UN official and co-author of a UN report detailing this problem, Henning Steinfeld, said "Livestock are one of the most significant contributors to today's most serious environmental problems". Livestock production occupies 70% of all land used for agriculture, or 30% of the land surface of the planet. It is one of the largest sources of greenhouse gases, responsible for 18% of the world's greenhouse gas emissions as measured in CO_2 equivalents.

By comparison, all transportation emits 13.5% of the CO_2. It produces 65% of human-related nitrous oxide (which has 296 times the global warming potential of CO_2) and 37% of all human-induced methane (which is 23 times as warming as CO_2. It also generates 64% of the ammonia, which contributes to acid rain and acidification of ecosystems.

Livestock expansion is cited as a key factor driving deforestation, in the Amazon basin 70% of previously forested area is now occupied by pastures and the remainder used for feedcrops. Through deforestation and land degradation, livestock is also driving reductions in biodiversity.

Land Transformation and Degradation

Land transformation, the use of land to yield goods and services, is the most substantial way humans alter the Earth's ecosystems, and is considered the driving force in the loss of biodiversity. Estimates of the amount of land transformed by humans vary from 39–50%. Land degradation, the long-term decline in ecosystem function and productivity, is estimated to be occurring on 24% of land worldwide, with cropland overrepresented.

The UN-FAO report cites land management as the driving factor behind degradation and reports that 1.5 billion people rely upon the degrading land. Degradation can be deforestation, desertification, soil erosion, mineral depletion, or chemical degradation (acidification and salinization).

Eutrophication

Eutrophication, excessive nutrients in aquatic ecosystems resulting in algal blooms and anoxia, leads to fish kills, loss of biodiversity, and renders water unfit for drinking and other industrial uses.

Excessive fertilization and manure application to cropland, as well as high livestock stocking densities cause nutrient (mainly nitrogen and phosphorus) runoff and leaching from agricultural land. These nutrients are major nonpoint pollutants contributing to eutrophication of aquatic ecosystems.

Pesticides

Pesticide use has increased since 1950 to 2.5 million tons annually worldwide, yet crop loss from pests has remained relatively constant. The World Health Organization estimated in 1992 that 3 million pesticide poisonings occur annually, causing 220,000 deaths. Pesticides select for pesticide resistance in the pest population, leading to a condition termed the 'pesticide treadmill' in which pest resistance warrants the development of a new pesticide.

An alternative argument is that the way to 'save the environment' and prevent famine is by using pesticides and intensive high yield farming, a view exemplified by a quote heading the Center for Global Food Issues website:

'Growing more per acre leaves more land for nature'. However, critics argue that a trade-off between the environment and a need for food is not inevitable, and that pesticides simply replace good agronomic practices such as crop rotation.

Climate Change

Climate change has the potential to affect agriculture through changes in temperature, rainfall (timing and quantity), CO_2, solar radiation and the interaction of these elements. Agriculture can both mitigate or worsen global warming. Some of the increase in CO_2 in the atmosphere comes from the decomposition of organic matter in the soil, and much of the methane emitted into the atmosphere is caused by the decomposition of organic matter in wet soils such as rice paddies.

Further, wet or anaerobic soils also lose nitrogen through denitrification, releasing the greenhouse gas nitric oxide. Changes in management can reduce the release of these greenhouse gases, and soil can further be used to sequester some of the CO_2 in the atmosphere.

International Economics and Agriculture

Differences in economic development, population density and culture mean that the farmers of the world operate under very different conditions. A US cotton farmer may receive US$230 in government subsidies per acre planted (in 2003), while farmers in Mali and other third-world countries do without.

When prices decline, the heavily subsidized US farmer is not forced to reduce his output, making it difficult for cotton prices to rebound, but his Mali counterpart may go broke in the meantime.

A livestock farmer in South Korea can calculate with a (highly subsidized) sales price of US$1300 for a calf produced. A South American Mercosur country rancher calculates with a calf's sales price of US$120–200 (both 2008 figures). With the former, scarcity and high cost of land is compensated with public subsidies, the latter compensates absence of subsidies with economics of scale and low cost of land.

In the Peoples Republic of China, a rural household's productive asset may be one hectare of farmland. In Brazil, Paraguay and other countries where local legislature allows such purchases, international investors buy thousands of hectares of farmland or raw land at prices of a few hundred US$ per hectare.

Energy and Agriculture

Since the 1940s, agricultural productivity has increased dramatically, due largely to the increased use of energy-intensive mechanization, fertilizers and pesticides.

The vast majority of this energy input comes from fossil fuel sources. Between 1950 and 1984, the Green Revolution transformed agriculture around the globe, with world grain production increasing by 250% as world population doubled. Modern agriculture's heavy reliance on petrochemicals and mechanization has raised concerns that oil shortages could increase costs and reduce agricultural output, causing food shortages.

Modern or industrialized agriculture is dependent on fossil fuels in two fundamental ways: 1) direct consumption on the farm and 2) indirect consumption to manufacture inputs used on the farm. Direct consumption includes the use of lubricants and fuels to operate farm vehicles and machinery; and use of gas, liquid propane, and electricity to power dryers, pumps, lights, heaters, and coolers. American farms directly consumed about 1.2 exajoules (1.1 quadrillion BTU) in 2002, or just over 1 percent of the nation's total energy.

Indirect consumption is mainly oil and natural gas used to manufacture fertilizers and pesticides, which accounted for 0.6 exajoules (0.6 quadrillion BTU) in 2002.

The energy used to manufacture farm machinery is also a form of indirect agricultural energy consumption, but it is not included in USDA estimates of U.S. agricultural energy use. Together, direct and indirect consumption by U.S. farms accounts for about 2 percent of the nation's energy use. Direct and indirect energy consumption by U.S. farms peaked in 1979, and has gradually declined over the past 30 years.

Food systems encompass not just agricultural production, but also off-farm processing, packaging, transporting, marketing, consumption, and disposal of food and food-related items. Agriculture accounts for less than one-fifth of food system energy use in the United States.

Oil shortages could impact this food supply. Some farmers using modern organic-farming methods have reported yields as high as those available from conventional farming without the use of synthetic fertilizers and pesticides. However, the reconditioning of soil to restore nutrients lost during the use of monoculture agriculture techniques made possible by petroleum-based technology takes time.

In 2007, higher incentives for farmers to grow non-food biofuel crops combined with other factors (such as over-development of former farm lands, rising transportation costs, climate change, growing consumer demand in China and India, and population growth) to cause food shortages in Asia, the Middle East, Africa, and Mexico, as well as rising food prices around the globe. As of December 2007, 37 countries faced food crises, and 20 had imposed some sort of food-price controls. Some of these shortages resulted in food riots and even deadly stampedes.

The biggest fossil fuel input to agriculture is the use of natural gas as a hydrogen source for the Haber-Bosch fertilizer-creation process. Natural gas is used because it is the cheapest currently available source of hydrogen. When oil production becomes so scarce that natural gas is used as a partial stopgap replacement, and hydrogen use in transportation increases, natural gas will become much more expensive.

If the Haber Process is unable to be commercialized using renewable energy (such as by electrolysis) or if other sources of hydrogen are not available to replace the Haber Process, in amounts sufficient to supply transportation and agricultural needs, this major source of fertilizer would either become extremely expensive or unavailable. This would either cause food shortages or dramatic rises in food prices.

Mitigation of Effects of Petroleum Shortages

One effect oil shortages could have on agriculture is a full return to organic agriculture. In light of peak-oil concerns, organic methods are more sustainable than contemporary practices because they use no petroleum-based pesticides, herbicides, or fertilizers. Some farmers using modern organic-farming methods have reported yields as high as those available from conventional farming.

Organic farming may however be more labour-intensive and would require a shift of the workforce from urban to rural areas. It has been suggested that rural communities might obtain fuel from the biochar and synfuel process, which uses agricultural *waste* to provide charcoal fertilizer, some fuel *and* food, instead of the normal food vs fuel debate.

As the synfuel would be used on-site, the process would be more efficient and might just provide enough fuel for a new organic-agriculture fusion.

It has been suggested that some transgenic plants may some day be developed which would allow for maintaining or increasing yields while requiring fewer fossil-fuel-derived inputs than conventional crops.

The possibility of success of these programs is questioned by ecologists and economists concerned with unsustainable GMO practices such as terminator seeds, and a January 2008 report shows that GMO practices "fail to deliver environmental, social and economic benefits."

While there has been some research on sustainability using GMO crops, at least one hyped and prominent multi-year attempt by Monsanto Company has been unsuccessful, though during the same period traditional breeding techniques yielded a more sustainable variety of the same crop. Additionally, a survey by the bio-tech industry of subsistence farmers in Africa to discover what GMO research would most benefit sustainable agriculture only identified non-transgenic issues as areas needing to be addressed. Nevertheless, some governments in Africa continue to view investments in new transgenic technologies as an essential component of efforts to improve sustainability.

Chapter 8

Nutritional Quality

Food Composition

- *Energy* for the metabolic and physiological functions of humans is derived from the chemical energy bound in food and its macronutrient constituents, i.e. *carbohydrates*, *fats*, *proteins* and *alcohol*, which act as substrates or fuels. After food is ingested, its chemical energy is released and converted into heat, mechanical and other forms of energy
- Food is made of one or several *macronutrients*, which are carbohydrates, fats, proteins and alcohol (e.g. rice contains 0.5% fat, 7% proteins, and 80% carbohydrates). Food contains also *micronutrients*, such as minerals (e.g. iron, calcium, etc.) and vitamins (e.g. vitamin A, B1 or C).
- Other components of the food are *water*, and sometimes *antinutritional factors*.
- Values for nutrients in food can *vary widely* and depend on such things as: the growing environment, the particular variety or breed, the method of processing, the storage temperature and time etc. The values given in the food specifications section are an indication of nutritional value.

Energy Provided by Macronutrients

Energy is needed for essential body functions (such as breathing), growth (especially during childhood), and physical activities (working and playing). Macronutrients provide different amounts of energy, expressed as kilocalories (Kcals). Fat provides approximately twice as much energy (9 kcals/g) as the same weight of protein or carbohydrate

(4 kcal/g). As stated above, more carbohydrate than fat is usually eaten in developing countries and, therefore, most food energy in the diet in these countries is derived from carbohydrate sources. The relative concentration of protein and fat in the diet is important and is expressed by the percentage of energy in the diet provided by either fat or protein. For example, if a diet provides 2,000 kcal, of which 200 kcal is provided by fat, the fat-energy percentage is 10.

Energy and Protein Requirements

The total amount of energy and protein needed by different individuals varies a great deal, depending primarily on the amount of physical activity but also on age, sex, body size and, to some extent, climate. Extra energy is needed during pregnancy and lactation. An average population, made from people of all age and gender, will need 2,100 kcal per day. Energy requirement will vary with age, sex, physical activity, climate, etc.

Energy Density Concept

The quantity of energy that a child can consume each day from gruels depends on the number of meals, quantities consumed during each meal and the energy density of the gruels. However, in many societies, the mothers, involved in multiple tasks, cannot prepare gruels more than twice a day.

In addition, babies cannot eat more than 30 to 40 ml of gruel per kilogram of body weight during each meal because of their reduced stomach capacity. In the case of gruels prepared from starch-based foods that have not undergone enzymatic or hydrothermal treatments (treatments involving water and temperature such as extrusion-cooking, drum-drying), the concentration of flour in the gruels is the main determinant of their energy density. The first approximation is that 4 kcal can be obtained from 1g of dry matter.

These gruels have a viscosity which increases very quickly according to their concentration in dry matter. Those who prepare these gruels are thus placed before the following dilemma: increase the proportion of flour with respect to water and obtain a gruel of very high viscosity, *difficult for children to swallow*, or prepare gruels of suitable consistency but of low energy density, and thus give more than three meals a day.

There is only one way to increase the quantities of energy consumed by children, wherever gruels are not given more than three times per day, only one way: increase the energy density of the gruels. For that, the flours must undergo enzymatic and/or hydrothermal treatments which modify the physicochemical properties of the starch. These treatments break the starch macromolecules, limit their swelling during cooking, and consequently change the viscosity of the gruels. It thus becomes possible to prepare gruels of higher energy density while maintaining an appropriate consistency.

Measurement of Gruels/Porridges Consistency

When a mother prepares a gruel for her child, it is undoubtedly the desired consistency of the gruel which guides her in the choice of the flour/water proportions. However, from these proportions depends the *energy density* of the gruel and, consequently, its' *nutritional value (i.e. its nutrient density).*

Thus, the higher the relative quantity of flour, the more the gruel will be nutritious, but also the more will it be thick, until the maximum consistency threshold accepted by the mother or the child is reached. The manufacture of infant flours permitting the preparation of gruels of energy density likely to satisfy the energy needs of the child, necessitates the application of technologies that reduce viscosity: addition of amylolytic enzymes, malting of a fraction of the raw materials, *extrusion-cooking*, *drum-drying*, etc...

In order to study the effectiveness of these technological treatments, an evaluation of consistency is necessary. The parameter generally measured is *viscosity*. But out of the laboratory, the realisation of reliable viscosity measurements is very difficult. In addition, the absence of recommendations for the adoption of standardised conditions of measurement invokes a great disparity in the measurement methods employed. There are many viscometers of different trade marks and characteristics and from one instrument to another, the measured viscosity on the same gruel can vary up to a factor of 10.

To evaluate the consistency of gruels, it is also possible to use a *Bostwick* consistometer. This empirical measurement, relatively simple to use in an enterprise, is less well conceived than the measurement of viscosity but could be better correlated with the sensory appreciation of mothers. Research studies are currently being carried out in the Laboratory for Tropical Nutrition, IRD, in order to better understand the relationship between viscosity, *Bostwick* flow and the sensory appreciation of the consistency of gruels.

The compartment that has a guillotine-type system is filled with a given volume (100 ml) of gruel. At t=0, the gruel is released and the consistency parameter obtained corresponds to the distance covered by the gruel front after 30 seconds of flow. The only control condition of the measurement is the gruel temperature: generally, a temperature close to that at which the gruel is consumed, that is to say approximately 45 °C is adopted.

Macronutrients

Protein

Proteins are made up of 'building blocks' called *amino acids*, composed of carbon, hydrogen, oxygen and nitrogen (amino group). Proteins from different food sources contain different amounts of amino acids. Proteins from animal origin, such as meat, milk and eggs, contain all *essential amino acids* in balanced amounts. Essential amino acids are those that the body cannot make itself and must therefore be eaten. In contrast, proteins of vegetable origin (e.g., cereals and pulses) contain on their own insufficient quantities of some of the essential amino acids. By combining different foods, however (e.g., cereals with beans), adequate levels of all amino acids can be obtained without requiring protein from animal sources. Proteins are required to build new tissue, particularly during the rapid growth period of infancy and early childhood, during pregnancy and nursing, and after infections or injuries. Excess protein is burned for energy.

The nutritional value of a protein food can be judged by its ability to provide both the quantity and number of essential amino acids needed by the body. Different food sources contain different groups of proteins, which are made up of different arrangements and amounts of amino acids. In general, proteins from animal sources are of greater nutritional value because they usually contain all the essential amino acids. Proteins from plant sources, such as cereals and vegetables, may be deficient in one or other of the essential amino acids. For example, the proteins obtained from wheat lack adequate quantities of one essential amino acid, and those from beans are deficient in another but the combiantion of cereal and pulses will provide a balanced diet. Because the content of amino acids is different in each food, when they are eaten together they complement each other and the mixture is of higher nutritional value than the separate foods, and is as good as animal protein. It is important, that a variety of different types of protein foods are eaten.

Cooking can alter the amino-acid composition of protein and this usually results in desirable flavour and browning development. Very little nutritional value is lost.

Protein Intake

WFP recommends that 10 - 12% of the energy of the ration is derived from protein.

For instance for a daily ration of 2,100 kcal;

- 10% = 210 kcal
- 210 kcal/4 kcal/g = 52.5g of protein per day
- Maize contains 8.5g of protein/100g.
- Beans (Cowpeas) contain 23.5g of protein/100g.
- A ration made from 400g of maize (400 x 8.5% = 34g) and 100g of Cowpeas (100 x 23.5% = 23.5g) will bring 57.5g of protein.

Fat

Fat is also known as lipid and is mainly present in food in a form called 'triglycerides'. Triglycerides consist of glycerol and three ('tri') fatty acids. The fatty acids can be mainly 'saturated' (in butter) or mainly 'polyunsaturated' (in margarine). There are also monounsaturated fatty acids, which occur in quantity in the triglycerides of olive oil and peanut oil. Fat is energy dense (9 kilocalories per gram), it does not mix with water, so that the food in which it is found tends to be more energy dense because of the relative lack of water. Eating fat in reasonable quantity is good for health. It must be remembered that the quality of the fat eaten as well as its quantity is important. Fat also confers texture and flavour to food, enhancing its palatability.

Types of Fats

There are several types of fats, the main ones are called triglycerides, cholesterol esters, and phospholipids. They contain fatty acids, which may be saturated, monounsaturated or polyunsaturated. Some polyunsaturated fatty acids are essential for humans because they cannot be made in our bodies. One group of essential fatty acids comes from plant sources; these fatty acids are found in considerable quantity in polyunsaturated margarine and in vegetable oils. We need about 1 to 2 per cent of our energy to come from this type of fatty acid.

Saturated fatty acids come mainly from ruminant animals, such as sheep and cattle, and from milk and dairy products of these animals. Monounsaturated fat does not increase the blood cholesterol level and may decrease it (olive oil and peanut oil provide monounsaturated fat). Polyunsaturated fat decreases it.

Fat Intake

WFP recommends that 17% of the energy of the ration is derived from fat.

For instance for a daily ration of 2,100 kcal;

- 17% = 357 kcal
- 357 kcal/ 9 kcal/g = 39.7 g of fat per day
- Maize contains 3g of fat per 100g. Beans contains 2g of fat per 100g.
- A ration made of 400g of maize (400 x 3% = 12g), 100g of beans (100 x 2% = 2g), and 25g of oil will cover the daily needs in fat (12 + 2 + 25 = 39g fat).

Carbohydrates

Carbohydrates can be divided into three main groups:

- Sugars
- Starches
- Dietary fibers.

Sugars and starches in food are sources of energy. Cellulose and some related substances are not used by our bodies as a significant source of energy. Nevertheless, these components are very important as, together with other indigestible substances, they constitute dietary fiber.

Sugars

The main sugars in food are sucrose, glucose, fructose, maltose and lactose. Sucrose is obtained from sugar cane (or sugarbeet) and is usually called 'sugar'. Besides providing energy, sugars also produce the sensation of sweetness. Each sugar contributes the same amount of energy (kilocalories) to our diet regardless of its sweetness. Different sugars are not equally sweet and the degree of sweetness of a food is often not a good indication of the amount of sugars present. For example, maltose is only half as sweet as sucrose.

Starch

Starch is the main form of carbohydrate in our food. It is found in a variety of cereals, vegetables and fruit, but mostly in cereals, flours and legumes (beans, peas). Starchy foods are usually cooked to improve digestibility and give a more desirable texture and flavour. During the ripening of fruit, starch is changed into sugars, which give sweetness to ripe fruits. In contrast to sugars, starch is often accompanied by significant amounts of other nutrients including dietary fiber. Starch has the same energy value as sugars.

Carbohydrate Intake

There is no specific dietary requirement for carbohydrate because energy can also be derived from protein, and fat. However, a diet that does not contain carbohydrate can lead to muscle breakdown and dehydration. This can be prevented by 50 to 100 grams of carbohydrate per day, but levels above this are desirable. Sources of complex carbohydrates, such as starch, are recommended as these often also provide necessary vitamins, elements (minerals) and dietary fiber.

Dietary Fibers

Dietary fibre is a term that refers to a group of food components that pass through the stomach and small intestine undigested and reach the large intestine virtually unchanged. Most other nutrients are digested and are being used in other parts of the body by this stage.

During its passage through the large intestine some components of dietary fiber are broken down to varying degrees and absorbed by the body; the remaining components are excreted in the faeces. The current attention being given to the role of dietary fiber in prevention of certain diseases is largely due to the observation that patterns of disease observed in Africa and Asia were different from those in Western countries. It was suggested that the dietary fiber content was associated with this difference. Although it is not yet proven, there is evidence to suggest that dietary fiber is beneficial for a good health.

Nutritionally Improved Food Feeds

The use of modern scientific practices such as biotechnology in agriculture has made it possible to introduce a specific characteristic in a particular grain that can improve its efficiency as a livestock feed. A wide range of options has been put forward by scientists and industry specialists as potential means of improving the nutritional composition of feed grains that would address the specific needs of different livestock industries.

In assessing research priorities in the area of feed grains quality improvement, there has been a lack of information on the economics of the various research options. In recognition of that knowledge gap, the Grains Research and Development Corporation (GRDC) funded a project, "Economic assessment of improving nutritional characteristics of feed grains (DAN331A)". The project was a collaborative one under the leadership of NSW Agriculture, involving the Australian Bureau of Agricultural and Resource Economics (ABARE) and ACE Livestock Consulting Pty Ltd.

That project aimed to provide for the first time a comprehensive set of information on the value of improving different characteristics of feed grains for animal nutrition, and information on who was likely to receive the benefits of the research. The objective of the analysis undertaken in this study was to assess those potential new feeds and determine the economic merit of research to develop those feeds.

A comprehensive set of options for new feed types has been evaluated, to establish the options with the highest priorities for research. In addition, to provide a benchmark for the value of the nutritional improvements, other forms of feed grains improvement were also assessed. The options analysed are classified as follows:

- Feeds involving change in protein content
- Feeds involving change in amino acid profile
- Feeds involving improvement in feed digestibility and efficiency
- Feeds involving reduction in anti-nutritional factors
- Feeds involving increase in yield
- New crop options.

The nutritional value of each of the new options was compared to the "standard" or unimproved feed grain. In some of the options, the nutritional quality of the grain can be changed without affecting its yield, and without any change in agronomic practices or the cost of production. In others, there were associated yield changes or changes in the level of inputs that would be needed to produce the nutritionally improved feed grain.

In assessing the relative benefits from alternative forms of improvement of nutrition of feed grains, the cost-reducing impacts of the different options have been analysed in a linear programming model that determines the least cost feed rations for the different livestock industries.

The aggregate model considers 43 feed ingredients and estimates the least cost feed rations for the 12 livestock industries simultaneously. The cost-reduction from the new feeds was identified for each livestock industry. Economic welfare analysis was then used to estimate the size and distribution of the benefits of research from the feed grains quality-improving research between the producers (including input suppliers such as grain producers) and the consumers (including processors and final consumers) of those livestock products. The analysis also identified which of the livestock industries were likely to receive the benefits from each of the new feeds.

All of the new feeds were analysed using the aggregate feed demand model, to give a comparative analysis of all the feeds. A selected subset of the new feeds was then analysed using ABARE's regional model. That analysis allowed some of the key potential new feeds to be examined in detail, while still being comparable to the full set of options. The analysis also reveals that the aggregate national analysis provides a valuable assessment of the overall value of the new feeds.

When the feeds were analysed to assess the economic benefits, a large number of the options were found to have small or very small returns that would not justify a significant research input. The analysis reveals that there are some opportunities to improve the productivity and competitiveness of Australia's livestock industries by improving the nutritional characteristics of some feed grains.

The feeds that provide the largest welfare benefits are: High oil lupins and Naked oats. The potential benefits from several other feeds are also sufficient to make them worthwhile research targets in the feed grains area, including: High oil sorghum, High protein lupins, Low arabinoxylan wheat, Hull-less barley, High oil oats, Low seed coat content barley, and High seed coat digestibility barley.

However, there are a large number of technically feasible potential new feeds that are not likely to produce sufficient benefits to make them a reasonable research target. Of the 25 feeds with improved nutritional characteristics that were analysed, 10 had total welfare benefits of less than $0.3 million per year and a further 6 less than $1.2 million per year.

Given the expected research costs, probabilities of success and the time lags involved in developing these feeds by plant breeding, it is unlikely that these options could be expected to provide a satisfactory rate of return on the research funds required. Research funds used for these projects could well be applied to more productive projects.

Several of those leading options for nutritional improvement had negative impacts on some livestock industries, so that none were able to provide universal benefits to all the industries included in the analysis.

As a result, different livestock industries would rank the potential new feeds in different ways, often markedly differently.

An alternative would be to aim for yield improvement rather than seek to improve the nutritional quality of the feeds. That direction for research funding would provide economic benefits of similar or greater size than from nutritional improvement, and the evidence from the analysis in this study is that those benefits may well be more evenly spread across the different industries.

Clearly, the selection of which new feeds to develop needs to be undertaken carefully, to ensure that scarce research and development funds are used to provide the best returns. The analysis in this study enables those feeds to be identified, so that research priorities for feed grains can be developed with improved knowledge of the economic consequences.

Nutritional Assessment Process for Biotechnology

Modern crop biotechnology uses new techniques to do what farmers and plant breeders have been doing for centuries: making small changes to the genetic make-up of plants to improve growing or eating characteristics.

The new knowledge and techniques that biotechnology has brought to plant breeding have significantly improved the precision and the speed of the breeding process, thereby significantly improving the efficiency with which scientists can develop new crop varieties for Asia.

Genetically modified foods are the most studied food products ever produced, but it is sometimes difficult for consumers to understand the exhaustive checking and testing processes that are applied to biotechnology-derived foods. Improving public knowledge of the safety assessment of biotechnology- derived foods is now thought to be an even greater priority and challenge than further development of the science and technology itself.

What people in Asia think about biotechnology AFIC's own consumer research found that ordinary people feel poorly informed about this topic area. They would like to know more, but in language and terms that are understandable and acceptable to scientists and non-scientists alike.

Indeed consumers have a right to know that their food is safe and nutritious. However, in order for consumer to feel confident that foods derived from modern biotechnology methods are safe to eat, they need some basic understanding of the safety tests and precautions that are applied the new varieties of crops and produce that are becoming available. New technologies, new language and knowledge As personal computers and the worldwide web emerged, a whole new family of words and phrases also emerged, which consumers have quickly adopted and become quite comfortable with. Similarly, modern biotechnology brings with it new language and new knowledge needs. One example of this is 'substantial equivalence'.

This is a term unfamiliar to most non-specialists, but which is fundamental to the principle of safety assessment for genetically modified foods. The Food and Agriculture Organization (FAO) and the World Health Organization (WHO) of the United Nations advocate the concept of 'substantial equivalence' as the most practical approach to address the safety evaluation of foods or food components derived by modern biotechnology.

In many Asian countries, the most readily understandable terms for foods derived from modern biotechnology breeding methods are food biotechnology or genetically modified foods. Abbreviations such as 'GM food' or GMOs are perceived as jargon. Such terminology may lead to confusion, miscommunication and even misinterpretation of the topic and related issues.

What is Substantial equivalence ? Substantial equivalence is based on the principle that, 'if a new food or food component is found to be substantially equivalent to an existing food or food component, it can be treated in the same manner with respect to safety'. For a foodstuff to be assessed as substantially equivalent to currently available products, the products in question is subjected to multiple tests and checks. These include molecular characterisation of the genetic modification, agronomic characterisation, nutritional assessment, toxicological assessment and safety assessment. For example, typical questions that have to be addressed are:

- Does the genetically modified food have a traditional counterpart that has a history of safe use?
- Has the concentration of any naturally occurring toxins or allergens in the food changed?
- Have the levels of key nutrients changed?
- Do new substances in the genetically modified food have a history of safe use?

- Has the food's digestibility been affected?
- Has the food been produced using accepted, established procedures?
- The overall goal of these tests is to determine whether the plant is substantially equivalent (in terms of chemical and nutritional composition and characteristics) to food derived from a conventional source that has a history of safe use.

A substantial equivalence evaluation focuses on the product rather than the process used to develop the product. If the new product is substantially equivalent to the conventional food or feed, then the product derived through biotechnology is considered to be as safe as the conventional counterpart.

If the food produced using biotechnology contains a new trait, which changes the levels of nutrients or anti-nutrients, such as a higher level of a vitamin or a lower level of an allergen, the assessment focuses on demonstrating the safety of the new trait. Researchers must prepare comprehensive data to support the safety and holesomeness of new crop varieties developed through biotechnology.

This process requires years of laboratory and field testing before a product can be brought to the market. This article is based on extracts from Food Biotechnology: a Communications Guide to Enhance Understanding. The Guide has been produced by the Asian Food Information Centre (AFIC) and the International Service for the Acquisition of Agri-biotech Applications (ISAAA). It is intended to provide anyone who needs to write or talk about biotechnology or who simply wants to understand the science and the issues related to this important topic area, with the necessary information resources.

The kit is designed to provide the most scientifically sound and up-to-date information about biotechnology products and processes, in language that both scientists and non-scientists can understand and agree upon.

The guide is available in English and other languages. Countries in Asia have a long history of producing foods using biotechnology including soy sauce, tempeh and natto (fermented soybeans), belacan (fermented shrimp paste), cincaluk (fermented shrimps), budu and ngoc nam (fermented fish sauce), tapai (fermented milk), toddy (fermented young flowers of palm) and sake. Foods such as pickles, vinegar, bread, yoghurt and cheese are also the products of biotechnology.

Modernadvances in biotechnology are not a panacea for all the challenges that Asia currently faces in providing its growing population with a varied, safe and high guality food supply. Nevertheless, biotechnology could contribute to this goal, through the development of crops that can give improved yields, require less pesticides, result in less enviromental degradation, better nutritional profiles, and better keeping and eating qualities to name but name but a few of the potential benefits. Some of the tests applied to improved crop varieties, to determine whether the plant is substantially equivalent (in terms of chemical and nutritional composition and characteristics) to food derived from a conventional source that has a history of safe use.

Safety assessment- Molecular characterisation - for new plant varieties produced through modern biotechnology, the source of the gene introduced into the plant is first identified. The transformation system used to insert the gene is defined as well as the number of copies of inserted genes and the integrity and stability of the genetic insert.

Agronomic traits - Usually the starting points for evaluating substantial equivalence. For example, in the case of potatoes, the traits commonly examined are yield, tuber size and distribution, dry matter content and disease resistance.

Nutritional assessment - Involves key nutrients including fats, proteins, carbohydrates and essential vitamins and minerals.

Toxicology assessment - Toxicants and anti-nutrients are compounds known to be naturally present in some crops that could have an impact on health if their levels increased. For example, solanine glycoalkaloids in potatoes or trypsin inhibitors in soybeans. The levels of antinutrients in crops produced through biotechnology are compared to conventionally produced varieties grown under comparable environmental and agronomic conditions.

Allergenicity - Genes from common allergenic foods are not used. Allergic responses to foods are almost always due to protein molecules in the food. Tests include examination of molecular structure, stability of protein in stomach and intestinal fluids and measurement of the amount of any new protein in the food.

The Concept of Substantial Equivalence, its Historical

Development and Current Use

For many years the practical difficulties of obtaining meaningful information on the safety of whole foods from conventional toxicology studies have been well recognized. This became particularly apparent from the vast number of animal feeding studies conducted to assess the safety of irradiated foods.

Animal studies are a major element in the safety assessment of many compounds such as pesticides, pharmaceuticals, industrial chemicals and food additives. In most cases however, the test substance is well characterised, of known purity, of no nutritional value and human exposure is generally low.

It is therefore relatively straightforward to feed such compounds to animals at a range of doses, some several orders of magnitude greater than the expected human exposure levels, in order to identify any potential adverse effects of importance to humans. In this way it is possible, in most cases, to determine levels of exposure at which adverse effects are not present, and so set safe upper limits by the application of appropriate safety factors.

By contrast, foods are complex mixtures of compounds characterised by wide variation in composition and nutritional value. Due to their bulk and effect on satiety they can usually only be fed to animals at low multiples of the amounts that might be present in the human diet.

In addition, a key factor to consider in conducting animal studies on foods is the nutritional value and balance of the diets used, to try to avoid the induction of adverse effects which are not related directly to the material itself. Picking up any potential adverse effects and relating these conclusively to an individual characteristic of the food can therefore be extremely difficult. Another consideration in deciding the need for animal studies is whether it is appropriate to subject experimental animals to such a study if it is unlikely to give rise to meaningful information.

In practice very few foods consumed today have been subject to any toxicological studies yet are generally accepted as being safe to eat. In developing a methodology for the safety assessment of new foods it was essential to establish a benchmark definition of safe food.

Development of a Safety Assessment Framework for GM Foods

Recognising that the development of GM foods was progressing rapidly the FAO and WHO convened an expert consultation in 1990 on the 'Assessment of Biotechnology in Food Production and Processing as Related to Food Safety'. The consultation recognised the limitations of traditional toxicological test methods when applied to whole foods and recommended that a more structured approach to safety assessment should be developed. The 1990 consultation identified the comparative principle whereby the food being assessed is compared with one that has an accepted level of safety, as being of considerable importance.

In 1993 the OECD published a report on the safety evaluation of foods derived by modern biotechnology. This report which was based on a number of intergovernmental consultations included a definition of safe food.

"Food is considered safe if there is reasonable certainty that no harm will result from its consumption under anticipated conditions. Historically, food prepared and used in traditional ways is considered safe on the basis of long term experience, even though it may naturally contain harmful substances. In principle, food is presumed to be safe unless a significant hazard has been identified."

This is not to say that many foods that are already widely consumed would not show adverse effects in animal studies if they could be fed at high enough doses. Equally given the many adverse effects that can be observed with existing foods it would be unreasonable to require a demonstration of absolute safety for novel foods.

The 1993 OECD report also expanded upon the comparative principle identified by the 1990 FAO/WHO consultation and formulated the concept of substantial equivalence.

The WHO and FAO refined the concept further at an expert consultation meeting held in Rome in 1996.

What is Substantial Equivalence

Substantial equivalence is not a substitute for a safety assessment, but a part of the assessment process. As such, it provides a useful framework for regulatory scientists. Underlying the concept is the requirement that any safety assessment should show that a genetically modified variety is as safe as its traditional counterparts, through a consideration of both intended and unintended effects. This involves consideration of a wide range of information, including agronomic properties, phenotypic changes and compositional data on critical nutrients and toxicants.

In the report of the 1996 expert consultation substantial equivalence was identified as being 'established by a demonstration that the characteristics assessed for the genetically modified organism, or the specific food product derived therefrom, are equivalent to the same characteristics of the conventional comparator. The levels and variation for characteristics in the genetically modified organism must be within the natural range of variation for those characteristics considered in the comparator and be based upon an appropriate analysis of data.'

Critical nutrients and toxicants are components of a particular crop known to be relevant to human health, as determined through our knowledge of the unmodified crop and its related species. Comparative assessment of these components and their potential for change as a result of genetic modification, together with a wide range of other information on agronomic, phenotypic and other properties, permits an assessment of the likelihood of unintended effects in a modified crop.

For example, where the level of expression of a particular gene is altered, this is likely to be reflected in other changes in either the crop's composition or appearance.

In the application of substantial equivalence, these key components are Considered on the basis of the long history of safe use of the traditional counterpart and any differences are identified. The defined differences are then the subject of safety assessment, which can include nutritional, toxicological and immunological testing as appropriate.

The concept of substantial equivalence has been used extensively as a tool in assessing the safety of GM foods. In comparing a GM food with a conventional counterpart, consideration is given to both intentional and unintentional effects. A wide range of information is used in these comparisons ranging from agronomic data such as crop height, yield, flowering pattern, disease resistance etc. through to compositional data on key nutrients and toxicants. In this context, key nutrients are those food components which may have a major impact on the total diet and include fats, proteins and carbohydrates as well as minerals and vitamins. The comparison can result in one of three conclusions:

- The GMO or food product obtained from it is substantially equivalent to a conventional counterpart.
- The GMO or food product obtained from it is substantially equivalent to a conventional counterpart except for a few clearly defined differences.

- The GMO or food product obtained from it is not substantially equivalent to a conventional counterpart - either because the differences cannot be defined or because there is no existing counterpart to compare it with.

Where a food can be demonstrated to be substantially equivalent, it is considered to be as safe as its counterpart and no further safety assessment is required.

Where there are clearly defined differences between the GM food and its conventional counterpart, the safety implications of the differences need to be fully assessed. Where a food is not substantially equivalent, it does not mean that the food is unsafe. However, there would be a need for extensive data to be provided to demonstrate its safety.

How Substantial Equivalence is Currently Used in Practice?

The concept of substantial equivalence has been integrated into safety assessment procedures used by regulatory authorities worldwide. In many countries e.g. the EU, Australia, Canada and the USA the safety assessment process has been formalised in a series of decision trees which guide regulators and potential applicants through the various stages of the comparative process.

At present each regulatory authority could, in theory, determine a unique set of components they wish to see analysed. In practice there is already broad agreement on key components, although work is progressing to develop international consensus on a core set of components on a crop by crop basis.

Although the concept has been interpreted in slightly different ways by various regulatory authorities, the overall approach to the safety assessment is very similar in all countries.

The area where there is perhaps the greatest scope for divergence is which of the three categories identified in paragraph 14 that a particular novel food is assigned to. This is largely down to differing interpretations as to what constitutes a difference from a conventional counterpart.

Whilst the foods derived from biotechnology which are on sale have all been assessed for safety using the substantial equivalence concept and the results endorsed by the respective Governments, there have been some published criticisms of substantial equivalence. Many of these criticisms are based on a misunderstanding of the concept. Nevertheless such criticisms provide a useful stimulus to ensure that safety assessment procedures are kept at the forefront of scientific knowledge.

The concept of substantial equivalence has been used by regulatory authorities worldwide for approximately ten years. In this time some 40 products have been assessed, the complexity of the genetic modifications has increased. For some products, particularly commodity crops such as soya, many millions of tonnes have been consumed with no evidence of any adverse health effects. The concept has stood the test of time, although it is important that it is kept under review to ensure it remains the most appropriate mechanism for assessing the nutritional and food safety implications of foods derived from biotechnology.

Chapter 9

Biochemistry of Food

Nutrition (also called nourishment or aliment) is the provision, to cells and organisms, of the materials necessary (in the form of food) to support life. Many common health problems can be prevented or alleviated with a healthy diet. The diet of an organism is what it eats, which is largely determined by the perceived palatability of foods. Dietitians are health professionals who specialize in human nutrition, meal planning, economics, and preparation. They are trained to provide safe, evidence-based dietary advice and management to individuals (in health and disease), as well as to institutions.

Overview

Nutrition science investigates the metabolic and physiological responses of the body to diet. With advances in the fields of molecular biology, biochemistry, and genetics, the study of nutrition is increasingly concerned with metabolism and metabolic pathways: the sequences of biochemical steps through which substances in living things change from one form to another.

Nitrogen is needed by animals to build proteins. Carnivore and herbivore diets vary in their source of nitrogen, which is a limiting nutrient for both. Herbivores consume plants to get nitrogen and carnivores consume other animals to obtain nitrogen. Nitrogen is a common element in the atmosphere but exists in a state that is not usable by most living organisms. Certain fungi and bacteria are able to convert atmospheric nitrogen into a form plants can adsorb and utilize. The human body contains chemical compounds, such as water, carbohydrates (sugar, starch, and fiber), amino acids (in proteins), fatty acids (in lipids), and nucleic acids (DNA and RNA).

These compounds in turn consist of elements such as carbon, hydrogen, oxygen, nitrogen, phosphorus, calcium, iron, zinc, magnesium, manganese, and so on. All of these chemical compounds and elements occur in various forms and combinations (e.g. hormones, vitamins, phospholipids, hydroxyapatite), both in the human body and in the plant and animal organisms that humans eat.

The human body consists of elements and compounds ingested, digested, absorbed, and circulated through the bloodstream to feed the cells of the body. Except in the unborn fetus, which receive processed nutrients from the mother, the digestive system is the first system involved in breaking down food prior to further digestion.

Digestive juices, excreted into the lumen of the gastrointestinal tract, break chemical bonds in ingested molecules, and modulate their conformations and energy states. Though some molecules are absorbed into the bloodstream unchanged, digestive processes release them from the matrix of foods. Unabsorbed matter, along with some waste products of metabolism, is eliminated from the body in the feces.

Studies of nutritional status must take into account the state of the body before and after experiments, as well as the chemical composition of the whole diet and of all material excreted and eliminated from the body (in urine and feces). Comparing the food to the waste can help determine the specific compounds and elements absorbed and metabolized in the body.

The effects of nutrients may only be discernible over an extended period, during which all food and waste must be analysed. The number of variables involved in such experiments is high, making nutritional studies time-consuming and expensive, which explains why the science of human nutrition is still slowly evolving.

In general, eating a wide variety of fresh, whole (unprocessed), foods has proven favourable for one's health compared to monotonous diets based on processed foods. In particular, the consumption of whole-plant foods slows digestion and allows better absorption, and a more favourable balance of essential nutrients per Calorie, resulting in better management of cell growth, maintenance, and mitosis (cell division), as well as better regulation of appetite and blood sugar. Regularly scheduled meals (every few hours) have also proven more wholesome than infrequent or haphazard ones, although a recent study has also linked more frequent meals with a higher risk of colon cancer in men.

Nutrients

There are six major classes of nutrients: carbohydrates, fats, minerals, protein, vitamins, and water. These nutrient classes can be categorized as either macronutrients (needed in relatively large amounts) or micronutrients (needed in smaller quantities). The macronutrients include carbohydrates, fats, protein, and water. The micronutrients are minerals and vitamins.

The macronutrients (excluding water) provide structural material (amino acids from which proteins are built, and lipids from which cell membranes and some signalling molecules are built), and energy.

Some of the structural material can be used to generate energy internally, and in either case it is measured in Joules or kilocalories (often called "Calories" and written with a capital *C* to distinguish them from little 'c' calories). Carbohydrates and proteins provide 17 kJ approximately (4 kcal) of energy per gram, while fats provide 37 kJ (9 kcal) per gram., though the net energy from either depends on such factors as absorption and digestive effort, which vary substantially from instance to instance. Vitamins, minerals, fiber, and water provide energy, but are required for other reasons. A third class of dietary material, fiber (i.e., non-digestible material such as cellulose), is also required, for both mechanical and biochemical reasons, although the exact reasons remain unclear.

Molecules of carbohydrates and fats consist of carbon, hydrogen, and oxygen atoms. Carbohydrates range from simple monosaccharides (glucose, fructose, galactose) to complex polysaccharides (starch). Fats are triglycerides, made of assorted fatty acid monomers bound to glycerol backbone. Some fatty acids, but not all, are essential in the diet: they cannot be synthesized in the body. Protein molecules contain nitrogen atoms in addition to carbon, oxygen, and hydrogen.

The fundamental components of protein are nitrogen-containing amino acids, some of which are essential in the sense that humans cannot make them internally. Some of the amino acids are convertible (with the expenditure of energy) to glucose and can be used for energy production just as ordinary glucose in a process known as gluconeogenesis. By breaking down existing protein, some glucose can be produced internally; the remaining amino acids are discarded, primarily as urea in urine. This occurs normally only during prolonged starvation.

Other micronutrients include antioxidants and phytochemicals, which are said to influence (or protect) some body systems. Their necessity is not as well established as in the case of, for instance, vitamins. Most foods contain a mix of some or all of the nutrient classes, together with other substances, such as toxins of various sorts. Some nutrients can be stored internally (e.g., the fat soluble vitamins), while others are required more or less continuously. Poor health can be caused by a lack of required nutrients or, in extreme cases, too much of a required nutrient. For example, both salt and water (both absolutely required) will cause illness or even death in excessive amounts.

Carbohydrates

Carbohydrates include sugars, starches and fiber. They constitute a large part of foods such as rice, noodles, bread, and other grain-based products. Carbohydrates may be classified chemically as monosaccharides, disaccharides, or polysaccharides depending on the number of monomer (saccharide or sugar) units they contain. Monosaccharides, disaccharides, and polysaccharides contain one, two, and three or more sugar units, respectively.

Polysaccharides are often referred to as *complex* carbohydrates because they consist of long, sometimes branched chains of single sugar units. Mono-and disaccharides are called simple carbohydrates. Dietary advice frequently but erroneously suggests that complex carbohydrates are superior to simple because they take longer to digest and absorb. Simple carbohydrates, on the other hand, are said to cause a spike in blood glucose levels rapidly after ingestion.

These traditional claims are false. In fact, many digestible polysaccharides are processed as rapidly as simple sugars in the human body. On the other hand some simple carbohydrates (fructose, for example) are processed in a different way and do not spike blood sugar. Thus the distinction between "complex" and "simple" does not predict the nutritional value or impact of carbohydrates. A better way of determining what effect particular foods may have on blood sugar and ultimately on health in general is the glycemic index.

Carbohydrates are not essential nutrients (with the likely exception of fiber), but are typically an important part of the human diet. While it would not be accurate to categorize all carbohydrates as "bad" nutritionally, some carbohydrate sources may well have deleterious effects on health, especially when consumed in large quantities. Highly processed carbohydrates (sugars and starches) as well as fructose consumed in large quantities have been implicated in negative health outcomes.

Fiber

Dietary fiber is a carbohydrate (or a polysaccharide) that is incompletely absorbed in humans and in some animals. Like all carbohydrates, when it is metabolized it can produce four Calories (kilocalories) of energy per gram. However, in most circumstances it accounts for less than that because of its limited absorption and digestibility. Dietary fiber consists mainly of cellulose, a large carbohydrate polymer that is indigestible because humans do not have the required enzymes to disassemble it. There are two subcategories: soluble and insoluble fiber. Whole grains, fruits (especially plums, prunes, and figs), and vegetables are good sources of dietary fiber. There are many health benefits of a high-fiber diet. Dietary fiber helps reduce the chance of gastrointestinal problems such as constipation and diarrhea by increasing the weight and size of stool and softening it. Insoluble fiber, found in whole-wheat flour, nuts and vegetables, especially stimulates peristalsis — the rhythmic muscular contractions of the intestines which move digesta along the digestive tract. Soluble fiber, found in oats, peas, beans, and many fruits, dissolves in water in the intestinal tract to produce a gel which slows the movement of food through the intestines. This may help lower blood glucose levels because it can slow the absorption of sugar. Additionally, fiber, perhaps especially that from whole grains, is thought to possibly help lessen insulin spikes, and therefore reduce the risk of type 2 diabetes. The link between increased fiber consumption and a decreased risk of colorectal cancer is still uncertain.

Fat

A molecule of dietary fat typically consists of several fatty acids (containing long chains of carbon and hydrogen atoms), bonded to a glycerol. They are typically found as triglycerides (three fatty acids attached to one glycerol backbone). Fats may be classified as saturated or unsaturated depending on the detailed structure of the fatty acids involved. Saturated fats have all of the carbon atoms in their fatty acid chains bonded to hydrogen atoms, whereas unsaturated fats have some of these carbon atoms double-bonded, so their molecules have relatively fewer hydrogen atoms than a saturated fatty acid of the same length.

Unsaturated fats may be further classified as monounsaturated (one double-bond) or polyunsaturated (many double-bonds). Furthermore, depending on the location of the double-bond in the fatty acid chain, unsaturated fatty acids are classified as omega-3 or omega-6 fatty acids. Trans fats are a type of unsaturated fat with *trans-*

isomer bonds; these are rare in nature and in foods from natural sources; they are typically created in an industrial process called (partial) hydrogenation. There are nine kilocalories in each gram of fat.

Saturated fats (typically from animal sources) have been a staple in many world cultures for millennia. Unsaturated fats (e.g., vegetable oil) are considered healthier (citation needed), while trans fats are to be avoided. Saturated and some trans fats are typically solid at room temperature (such as butter or lard), while unsaturated fats are typically liquids (such as olive oil or flaxseed oil). Trans fats are very rare in nature, and have been shown to be highly detrimental to human health, but have properties useful in the food processing industry, such as rancidity resistance.

Essential Fatty Acids

Most fatty acids are non-essential, meaning the body can produce them as needed, generally from other fatty acids and always by expending energy to do so. However, in humans, at least two fatty acids are essential and must be included in the diet. An appropriate balance of essential fatty acids—omega-3 and omega-6 fatty acids—seems also important for health, although definitive experimental demonstration has been elusive. Both of these "omega" long-chain polyunsaturated fatty acids are substrates for a class of eicosanoids known as prostaglandins, which have roles throughout the human body. They are hormones, in some respects.

The omega-3 eicosapentaenoic acid (EPA), which can be made in the human body from the omega-3 essential fatty acid alpha-linolenic acid (LNA), or taken in through marine food sources, serves as a building block for series 3 prostaglandins (e.g. weakly inflammatory PGE3). The omega-6 dihomo-gamma-linolenic acid (DGLA) serves as a building block for series 1 prostaglandins (e.g. anti-inflammatory PGE1), whereas arachidonic acid (AA) serves as a building block for series 2 prostaglandins (e.g. pro-inflammatory PGE 2). Both DGLA and AA can be made from the omega-6 linoleic acid (LA) in the human body, or can be taken in directly through food. An appropriately balanced intake of omega-3 and omega-6 partly determines the relative production of different prostaglandins, which is one reason why a balance between omega-3 and omega-6 is believed important for cardiovascular health. In industrialized societies, people typically consume large amounts of processed vegetable oils, which have reduced amounts of the essential fatty acids along with too much of omega-6 fatty acids relative to omega-3 fatty acids.

The conversion rate of omega-6 DGLA to AA largely determines the production of the prostaglandins PGE1 and PGE2. Omega-3 EPA prevents AA from being released from membranes, thereby skewing prostaglandin balance away from pro-inflammatory PGE2 (made from AA) toward anti-inflammatory PGE1 (made from DGLA). Moreover, the conversion (desaturation) of DGLA to AA is controlled by the enzyme delta-5-desaturase, which in turn is controlled by hormones such as insulin (up-regulation) and glucagon (down-regulation). The amount and type of carbohydrates consumed, along with some types of amino acid, can influence processes involving insulin, glucagon, and other hormones; therefore the ratio of omega-3 versus omega-6 has wide effects on general health, and specific effects on immune function and inflammation, and mitosis (i.e. cell division).

Protein

Proteins are the basis of many animal body structures (e.g. muscles, skin, and hair). They also form the enzymes that control chemical reactions throughout the body. Each molecule is composed of amino acids, which are characterized by inclusion of nitrogen and sometimes sulphur (these components are responsible for the distinctive smell of burning protein, such as the keratin in hair). The body requires amino acids to produce new proteins (protein retention) and to replace damaged proteins (maintenance). As there is no protein or amino acid storage provision, amino acids must be present in the diet. Excess amino acids are discarded, typically in the urine.

For all animals, some amino acids are *essential* (an animal cannot produce them internally) and some are *non-essential* (the animal can produce them from other nitrogen-containing compounds). Twenty-one proteinogenic amino acids are found in the human body, along with non-proteinogenic amino acids (e.g. gamma-aminobutyric acid).

Ten of the proteinogenic amino acids are essential and, therefore, must be included in the diet. A diet that contains adequate amounts of amino acids (especially those that are essential) is particularly important in some situations: during early development and maturation, pregnancy, lactation, or injury (a burn, for instance). A *complete* protein source contains all the essential amino acids; an *incomplete* protein source lacks one or more of the essential amino acids.

Sources of dietary protein include meats, tofu and other soy-products, eggs, legumes, and dairy products such as milk and cheese. Excess amino acids from protein can be converted into glucose and used for fuel through a process called gluconeogenesis. The amino acids remaining after such conversion are discarded.

Minerals

Dietary minerals are the chemical elements required by living organisms, other than the four elements carbon, hydrogen, nitrogen, and oxygen that are present in nearly all organic molecules. The term "mineral" is archaic, since the intent is to describe simply the less common elements in the diet. Some are heavier than the four just mentioned, including several metals, which often occur as ions in the body. Some dietitians recommend that these be supplied from foods in which they occur naturally, or at least as complex compounds, or sometimes even from natural inorganic sources (such as calcium carbonate from ground oyster shells). Some minerals are absorbed much more readily in the ionic forms found in such sources. On the other hand, minerals are often artificially added to the diet as supplements; the most famous is likely iodine in iodized salt which prevents goiter.

Macrominerals

Many elements are essential in relative quantity; they are usually called "bulk minerals". Some are structural, but many play a role as electrolytes. Elements with recommended dietary allowance (RDA) greater than 200 mg/day are, in alphabetical order (with informal or folk-medicine perspectives in parentheses):

- Calcium, a common electrolyte, but also needed structurally (for muscle and digestive system health, bone strength, some forms neutralize acidity, may help clear toxins, provides signalling ions for nerve and membrane functions)
- Chlorine as chloride ions; very common electrolyte;
- Magnesium, required for processing ATP and related reactions (builds bone, causes strong peristalsis, increases flexibility, increases alkalinity)
- Phosphorus, required component of bones; essential for energy processing
- Potassium, a very common electrolyte (heart and nerve health)
- Sodium, a very common electrolyte; not generally found in dietary supplements, despite being needed in large quantities, because the ion is very common in food: typically as sodium chloride, or common salt. Excessive sodium consumption can deplete calcium and magnesium, leading to high blood pressure

and osteoporosis (Note: Some sources suggest high blood pressure is due to high water retention per osmosis).

- Sulfur, for three essential amino acids and therefore many proteins (skin, hair, nails, liver, and pancreas). Sulfur is not consumed alone, but in the form of sulfur-containing amino acids

Trace Minerals

Many elements are required in trace amounts, usually because they play a catalytic role in enzymes. Some trace mineral elements (RDA < 200 mg/day) are, in alphabetical order:

- Cobalt required for biosynthesis of vitamin B_{12} family of coenzymes. Animals cannot biosynthesize B_{12}, and must obtain this cobalt-containing vitamin in the diet
- Copper required component of many redox enzymes, including cytochrome c oxidase
- Chromium required for sugar metabolism
- Fluoride required for maintenance of bone structure
- Iodine required not only for the biosynthesis of thyroxine, but probably, for other important organs as breast, stomach, salivary glands, thymus etc.; for this reason iodine is needed in larger quantities than others in this list, and sometimes classified with the macrominerals
- Iron required for many enzymes, and for hemoglobin and some other proteins
- Manganese (processing of oxygen)
- Molybdenum required for xanthine oxidase and related oxidases
- Nickel present in urease
- Selenium required for peroxidase (antioxidant proteins)
- Vanadium (Speculative: there is no established RDA for vanadium. No specific biochemical function has been identified for it in humans, although vanadium is required for some lower organisms.)
- Zinc required for several enzymes such as carboxypeptidase, liver alcohol dehydrogenase, and carbonic anhydrase

Vitamins

Some vitamins are recognized as essential nutrients, necessary in the diet for good health. Vitamin D is the exception: it can be synthesized in the skin, in the presence of UVB radiation. Certain vitamin-like compounds that are recommended in the diet, such as carnitine, are thought useful for survival and health, but these are not "essential" dietary nutrients because the human body has some capacity to produce them from other compounds.

Moreover, thousands of different phytochemicals have recently been discovered in food (particularly in fresh vegetables), which may have desirable properties including antioxidant activity; however, experimental demonstration has been suggestive but inconclusive. Other essential nutrients that are not classified as vitamins include essential amino acids, choline, essential fatty acids, and the minerals.

Vitamin deficiencies may result in disease conditions, including goitre, scurvy, osteoporosis, impaired immune system, disorders of cell metabolism, certain forms of cancer, symptoms of premature aging, and poor psychological health (including eating disorders), among many others. Excess levels of some vitamins are also dangerous to health (notably vitamin A), and for at least one vitamin, B6, toxicity begins at levels not far above the required amount. Deficient or excess levels of minerals can also have serious health consequences.

Water

It is not fully clear how much water intake is needed by healthy people, although some assert that 6–8 glasses of water daily is the minimum to maintain proper hydration. The notion that a person should consume eight glasses of water per day cannot be traced to a credible scientific source. The effect of, greater or lesser, water intake on weight loss and on constipation is also still unclear. The original water intake recommendation in 1945 by the Food and Nutrition Board of the National Research Council read: "An ordinary standard for diverse persons is 1 milliliter for each calorie of food. Most of this quantity is contained in prepared foods."

The latest dietary reference intake report by the United States National Research Council recommended, generally, (including food sources): 2.7 litres of water total for women and 3.7 litres for men. Specifically, pregnant and breastfeeding women need additional fluids to stay hydrated. According to the Institute of Medicine—who recommend that, on average, women consume 2.2 litres and men 3.0 litres—this is recommended to be 2.4 litres (approx. 9 cups) for pregnant women and 3 litres (approx. 12.5 cups) for breastfeeding women because an especially large amount of fluid is lost during nursing.

For those who have healthy kidneys, it is somewhat difficult to drink too much water, but (especially in warm humid weather and while exercising) it is dangerous to drink too little. People can drink far more water than necessary while exercising, however, putting them at risk of water intoxication, which can be fatal. In particular, large amounts of de-ionized water are dangerous.

Normally, about 20 percent of water intake comes in food, while the rest comes from drinking water and assorted beverages (caffeinated included). Water is excreted from the body in multiple forms; including urine and feces, sweating, and by water vapour in the exhaled breath.

Antioxidants

As cellular metabolism/energy production requires oxygen, potentially damaging (e.g. mutation causing) compounds known as free radicals can form. Most of these are oxidizers (i.e. acceptors of electrons) and some react very strongly. For the continued normal cellular maintenance, growth, and division, these free radicals must be sufficiently neutralized by antioxidant compounds.

Some are produced by the human body with adequate precursors (glutathione, Vitamin C), and those the body cannot produce may only be obtained in the diet via direct sources (Vitamin C in humans, Vitamin A, Vitamin K) or produced by the body from other compounds (Beta-carotene converted to Vitamin A by the body, Vitamin D synthesized from cholesterol by sunlight).

Phytochemicals (*Section Below*) and their subgroup, polyphenols, make up the majority of antioxidants; about 4,000 are known. Different antioxidants are now known to function in a cooperative network. For example, Vitamin C can reactivate free radical-containing glutathione or Vitamin E by accepting the free radical itself. Some antioxidants are more effective than others at neutralizing different free radicals.

Some cannot neutralize certain free radicals. Some cannot be present in certain areas of free radical development (Vitamin A is fat-soluble and protects fat areas, Vitamin C is water soluble and protects those areas). When interacting with a free radical, some antioxidants produce a different free radical compound that is less dangerous or more dangerous than the previous compound. Having a variety of antioxidants allows any byproducts to be safely dealt with by more efficient antioxidants in neutralizing a free radical's butterfly effect.

Although initial studies suggested that antioxidant supplements might promote health, later large clinical trials did not detect any benefit and suggested instead that excess supplementation may be harmful.

Phytochemicals

A growing area of interest is the effect upon human health of trace chemicals, collectively called phytochemicals. These nutrients are typically found in edible plants, especially colourful fruits and vegetables, but also other organisms including seafood, algae, and fungi. The effects of phytochemicals increasingly survive rigorous testing by prominent health organizations. One of the principal classes of phytochemicals are polyphenol antioxidants, chemicals that are known to provide certain health benefits to the cardiovascular system and immune system. These chemicals are known to down-regulate the formation of reactive oxygen species, key chemicals in cardiovascular disease.

Perhaps the most rigorously tested phytochemical is zeaxanthin, a yellow-pigmented carotenoid present in many yellow and orange fruits and vegetables. Repeated studies have shown a strong correlation between ingestion of zeaxanthin and the prevention and treatment of age-related macular degeneration (AMD). Less rigorous studies have proposed a correlation between zeaxanthin intake and cataracts. A second carotenoid, lutein, has also been shown to lower the risk of contracting AMD. Both compounds have been observed to collect in the retina when ingested orally, and they serve to protect the rods and cones against the destructive effects of light.

Another carotenoid, beta-cryptoxanthin, appears to protect against chronic joint inflammatory diseases, such as arthritis. While the association between serum blood levels of beta-cryptoxanthin and substantially decreased joint disease has been established, neither a convincing mechanism for such protection nor a cause-and-effect have been rigorously studied. Similarly, a red phytochemical, lycopene, has substantial credible evidence of negative association with development of prostate cancer.

As indicated above, some of the correlations between the ingestion of certain phytochemicals and the prevention of disease are, in some cases, enormous in magnitude. Yet, even when the evidence is obtained, translating it to practical dietary advice can be difficult and counter-intuitive. Lutein, for example, occurs in many yellow and orange fruits and vegetables and protects the eyes against various diseases. However, it does not protect the eye nearly as well as zeaxanthin, and the presence of lutein in the retina will prevent zeaxanthin uptake.

Additionally, evidence has shown that the lutein present in egg yolk is more readily absorbed than the lutein from vegetable sources, possibly because of fat solubility. At the most basic level, the question "should you eat eggs?" is complex to the point of dismay, including misperceptions about the health effects of cholesterol in egg yolk, and its saturated fat content.

As another example, lycopene is prevalent in tomatoes (and actually is the chemical that gives tomatoes their red colour). It is more highly concentrated, however, in processed tomato products such as commercial pasta sauce, or tomato soup, than in fresh "healthy" tomatoes. Yet, such sauces tend to have high amounts of salt, sugar, other substances a person may wish or even need to avoid.

Intestinal Bacterial Flora

It is now also known that animal intestines contain a large population of gut flora. In humans, these include species such as *Bacteroides*, *L. acidophilus*, and *E. coli*, among many others. They are essential to digestion, and are also affected by the food we eat. Bacteria in the gut perform many important functions for humans, including breaking down and aiding in the absorption of otherwise indigestible food; stimulating cell growth; repressing the growth of harmful bacteria, training the immune system to respond only to pathogens; producing vitamin B_{12}, and defending against some infectious diseases.

Advice and Guidance

Governmental Policies

In the US, dietitians are registered (RD) or licensed (LD) with the Commission for Dietetic Registration and the American Dietetic Association, and are only able to use the title "dietitian," as described by the business and professions codes of each respective state, when they have met specific educational and experiential prerequisites and passed a national registration or licensure examination, respectively.

Anyone may call themselves a nutritionist, including unqualified dietitians, as this term is unregulated. Some states, such as the State of Florida, have begun to include the title "nutritionist" in state licensure requirements.

Most governments provide guidance on nutrition, and some also impose mandatory disclosure/labelling requirements for processed food manufacturers and restaurants to assist consumers in complying with such guidance.

In the US, nutritional standards and recommendations are established jointly by the US Department of Agriculture and US Department of Health and Human Services. Dietary and physical activity guidelines from the USDA are presented in the concept of a food pyramid, which superseded the Four Food Groups. The Senate committee currently responsible for oversight of the USDA is the *Agriculture, Nutrition and Forestry Committee*. Committee hearings are often televised on C-SPAN as seen here.

The U.S. Department of Health and Human Services provides a sample week-long menu which fulfills the nutritional recommendations of the government. Canada's Food Guide is another governmental recommendation.

Teaching

Nutrition is taught in schools in many countries. In England and Wales the Personal and Social Education and Food Technology curricula include nutrition, stressing the importance of a balanced diet and teaching how to read nutrition labels on packaging. In many schools a Nutrition class will fall within the Family and Consumer Science or Health departments. In some American schools, students are required to take a certain number of FCS or Health related classes. Nutrition is offered at many schools, and if it is not a class of its own, nutrition is included in other FCS or Health classes such as: Life Skills, Independent Living, Single Survival, Freshmen Connection, Health etc. In many Nutrition classes, students learn about the food groups, the food pyramid, Daily Recommended Allowances, calories, vitamins, minerals, malnutrition, physical activity, healthy food choices and how to live a healthy life.

A 1985 US National Research Council report entitled *Nutrition Education in US Medical Schools* concluded that nutrition education in medical schools was inadequate. Only 20% of the schools surveyed taught nutrition as a separate, required course. A 2006 survey found that this number had risen to 30%.

Healthy Diets

The United Healthcare/Pacificare nutrition guideline recommends the following foods per day to enhance your health (based on a 2,000 calorie diet)

- Grains: 6 oz per day 'Make half your grains whole'
- Vegetables: 2.5 cups per day 'Vary your veggies'
- Fruits: 2 cups per day 'Focus on fruits'

- Milk: 3 cups per day 'Get your calcium-rich foods'
- Meat and beans: 5.5 oz per day 'Go lean with protein'.

A *National Geographic* cover article from November, 2005, entitled *The Secrets of Living Longer*, looks into why certain populations have high life spans. The article is a lifestyle survey of three populations, Sardinians, Okinawans, and Adventists, who generally display longevity and "suffer a fraction of the diseases that commonly kill people in other parts of the developed world, and enjoy more healthy years of life." In sum, they offer three sets of 'best practices' to emulate. The rest is up to you. In common with all three groups is to follow a low-calorie diet that consist mainly of "fruits, vegetables, and whole grains."

The *National Geographic* article noted that an NIH funded study of 34,000 Seventh-day Adventists between 1976 and 1988 "...found that the Adventists' habit of consuming beans, soy milk, tomatoes, and other fruits lowered their risk of developing certain cancers. It also suggested that eating whole grain bread, drinking five glasses of water a day, and, most surprisingly, consuming four servings of nuts a week reduced their risk of heart disease."

The French "Paradox"

The French paradox is the observation that the French suffer a relatively low incidence of coronary heart disease, despite having a diet relatively rich in saturated fats. A number of explanations have been suggested:

- Saturated fat consumption does not cause heart disease
- Reduced consumption of processed carbohydrate and other junk foods.
- Regular consumption of red wine.
- More active lifestyles involving plenty of daily exercise, especially walking; the French are much less dependent on cars than Americans are.
- Higher consumption of artificially produced trans-fats by Americans, which has been shown to have greater lipoprotein effects per gram than saturated fat.

On the other hand, statistics collected by the World Health Organization from 1990-2000 show that the incidence of heart disease in France may have been underestimated and, in fact, may be similar to that of neighbouring countries.

Sports Nutrition

Sports nutrition is the study and practice of nutrition and diet as it relates to athletic performance. It is concerned with the type and quantity of fluid and food taken by an athlete, and deals with nutrients such as vitamins, minerals, supplements and organic substances such as carbohydrates, proteins and fats. Although an important part of many sports training regimens, it is most commonly considered in strength sports (such as weight lifting and bodybuilding) and endurance sports (for example cycling, running, triathlon).

History

Ancient

From the time when the ancient Greeks and Romans started the Olympic games, the athletes had their own special regimen for great performance which included diet and nutrition. For example, Milo of Croton, the wrestler with legendary strength who won five successive Olympic Games from 532 to 516 B.C., ate "9 kilograms (20 lb) of meat, 9 kilograms (20 lb) of bread and 8.5 litres (18 US pt) of wine a day," according to Athenaeus and Pausanias (Grandjean). Alcohol was commonly drunk as an ergogenic aid to increase performance in the Olympics through the early 1900s.

Modern

The view of sports nutrition today has much evolved from the ancient Olympic gladiators' meal plan. Scientists are continually interested in learning more about this abiding subject. Research shows that the coupling of exercise and proper diet is what produces a healthy lifestyle that can maintain the "prevention/management of [chronic diseases such as] noninsulin-independent diabetes, hypertension, coronary heart disease, osteoporosis, obesity, mental health, colon cancers, stroke and back injury."

In 2008, US News reported that 65% of Americans exercised regularly by working out, playing sports, and/or other physical activities, thus the importance of proper nutrition is of great interest to athletes and exercisers for optimal performance and long term benefits.

In addition to diet, social and cultural influences, lifestyle habits, motivation and training determine successful athletic performance (Maughan). However, states Maughan, "without proper nutrition, the full potential of the athlete will not be realized, because performance will not be at its peak, training levels may not be sustained, recovery from injury will be slower, and the athlete may become more susceptible to injury and infection." Understanding sports nutrition leads to optimal athletic performance and lifetime health benefits and can thus be evaluated by the intake of certain nutrients and supplements when exercising, by learning the way the body utilizes these materials and how these practices complement future diet and exercise of the individual.

Goals

Sports nutrition has many goals to enhance performance:

First, it improves performance by improving body composition, which increases speed, quickness, mobility, and strength.

Second, it will help the speed of recovery, which will in turn create more capacity for practicing and competition as the body is becoming more fit and adjusted to the coupling of the good nutrition incorporated into the workout regimen.

Third, it will allow one to increase energy for both practice and competition, which will definitely help one's performance. Strategic diet will also increase immunity, allowing one to stay healthy and be able to continue and intensify practice and training. Most importantly, it will improve your overall health as proper health is essential to all aspects of life.

Supplements

Many athletes consider taking dietary supplements because they are looking for the "magic ingredient" to increase performance. In the extreme case of performance-enhancing supplements, athletes (particularly bodybuilders) may choose to use illegal substances such as anabolic steroids, compounds which are related to the hormone testosterone, which can quickly build mass and strength, but have many adverse effects such as high blood pressure and negative gender specific effects. Blood doping, another illegal ergogenic, was discovered in the 1940s when it was used by World War II pilots.

Dietary protein began to be consumed in the 1940s and muscle building results were found in resistance and strength training athletes. Protein intake is a part of the nutrient requirements for the regular athlete and is an important component of exercise training, because it can also aid in performance and recovery. Dietary protein intake for well-trained athletes should occur before, during and after physical activity as it is advantageous in gaining muscle mass and strength. However, if too much protein and amino acid supplements are consumed (especially by the Average Joe exerciser), it can be more harmful than beneficial; health risks include: "dehydration, gout, calcium loss, liver, and renal damage [and] gastrointestinal side effects include diarrhea, bloating, and water loss" (Lawerence). A bountiful protein diet must be paired with a healthy, well-rounded meal plan and regular resistance exercise. Yet, characteristics such as the type of exercise, intensity, duration, the carbohydrate values of diet, the individual's sex and age and also the amount of background training and training environment.

Creatine may be helpful for well-trained athletes to increase exercise performance and strength in concordance with their dietary regimen. Also, the substance glutamine, found in whey protein supplements, is the most abundant free amino acid found in the human body. For well-trained and well-nourished athletes it is considered that glutamine may have a possible role in stimulated anabolic processes such muscle glycogen and protein synthesis. Other popular supplements studies done include androstenedione, chromium, and ephedra. The findings show that there are no substantial benefits from the extra intake of these supplements, yet higher health risks and costs.

High energy supplements have shown to increase the performance of physical activity. A study done at the University of Texas saw a 4.7% increase of performance in 83% of participants after drinking Red Bull Energy Drink which was more intense than the compared placebo. The energy drink most dominantly increased the epinephrine and noreprinephrine (adrenaline and its precursor) levels and beta-endorphins in the blood than before consumption. Caffeine, carbohydrates and Vitamin B are factors that may have favoured performance increase with no change in perceived exertion.

Caffeine has been known since the 1900s and became popularly used since the 1970s when its power of masking fatigue became highly recognized. Similarly, the caffeine found in energy drinks shows an increased reaction performance and increased good feelings of energy, focus and alterness in quickness and reaction anaerobic power tests. In other words, consuming an energy drink with caffeine increases short time/rapid exercise performance (like short full-speed sprints and heavy power weight lifting.)

Post-exercise nutrition is just as important, if not more important than pre-exercise nutrition as it pertains to recovery. Traditionally, sports drinks such as Gatorade and Powerade, are consumed during and after exercise because they effectively rehydrate the body by refueling the body with minerals and electrolytes. Gatorade was founded in the 1960s, when the University of Florida, Gainesville Gators improved their performance with "Gator Aid." A drink was made of glucose and sucrose in water and helped the football players' performance. And by the 1970s, many other sports drinks of its kind had been manufactured. However, sports drinks lack protein.

New studies in 2008 have found milk, especially skim milk and chocolate milk may be the new sports drink, as milk leads to protein the synthesis which boosts net muscle protein balance. Milk naturally contains many electrolytes, nutrients and other properties that help to make it a great post-exercise beverage. When compared to plain water or sports drinks, research suggests that chocolate milk is more effective at replacing fluids lost as sweat and maintaining normal body fluid levels. Athletes drinking chocolate milk following exercise-induced dehydration had fluid levels about 2 percent higher (on initial body mass) than those using other post-exercise recovery beverages, allowing for prolonged performance, especially in repeated bouts of exercise or training..

Factors Influencing Nutritional Requirements

Differing conditions and objectives suggest the need for athletes to ensure that their nutritional approach is appropriate for their situation. Factors that may affect an athlete's nutritional needs include type of activity (aerobic vs. anaerobic), gender, weight, height, body mass index, workout or activity stage (pre-workout, intra-workout, recovery), and time of day (e.g. some nutrients are utilized by the body more effectively during sleep than while awake).

Anaerobic Exercise

During anaerobic exercise, the process of glycolysis breaks down the sugars from carbohydrates for energy without the use of oxygen. This type of exercise occurs in physical activity such as power sprints, strength resistances and quick explosive movement where the muscles are being used for power and speed, with short time energy use. After this type of exercise, there is a need to refill glycogen storage sites in the body (the long simple sugar chains in the body that store energy), although they are not likely fully depleted. To compensate for this glycogen reduction, athletes will often take in a large amount of carbohydrates in the period immediately following exercise. Typically, high glycemic index carbohydrates are preferred for their ability to rapidly raise blood glucose levels.

For the purpose of protein synthesis, protein or individual amino acids are ingested as well. Branched-chain amino acids are important since they are most responsible for protein synthesis. Often in the continuation of this anaerobic exercise, the product from this metabolic mechanism builds up in what is called lactic acid fermentation. Lactate is produced more quickly than it is being removed and it serves to regenerate NAD^+ to the cells where it's needed. During intense exercise when oxygen is not being used, a high amount of ATP is produced and pH levels fall causing acidosis or more specifically lactic acidosis. Lactic acid build up can be treated by staying well-hydrated throughout and especially after the workout, having good cool down routine and good post-workout stretching.

Protein

Protein is an important component of every cell in the body. Hair and nails are mostly made of protein. The body uses protein to build and repair tissues. In addition, protein is used to make hormones and other chemicals in the body. Protein is also an important building block of bones, muscles, cartilage, skin, and blood.

The protein requirement for each individual differs, as do opinions about whether and to what extent physically active people require more protein. The 2005 Recommended Dietary Allowances (RDA), aimed at the general healthy adult population, provide for an intake of 0.8 - 1 grams of protein per kilogram of body weight (according to the BMI formula), with the review panel stating that "no additional dietary protein is suggested for healthy adults undertaking resistance or endurance exercise". Conversely, Di Pasquale (2008), citing recent studies, recommends a minimum protein intake of 2.2 g/kg "for anyone involved in competitive or intense recreational sports who wants to maximize lean body mass but does not wish to gain weight".

Water

Water is one of the most important nutrients in the sports diet. It helps eliminate food waste products in the body, regulates body temperature during activity and helps with digestion. Maintaining hydration during periods of physical exertion is key to peak performance. While drinking too much water during activities can lead to physical discomfort, dehydration in excess of 2% of body mass (by weight) markedly hinders athletic performance. Additional carbohydrates and protein before, during, and after exercise increase time to exhaustion as well as speed recovery. The amount of water needed is based on work performed, lean body mass, and environmental factors, especially ambient temperature and humidity.

Carbohydrates

The main fuel used by the body during exercise is carbohydrates, which are stored in muscle as glycogen—a form of sugar. During exercise, muscle glycogen reserves can be used up, especially when activities last longer than 90 min. Because the amount of glycogen stored in the body is limited, it is important for athletes to replace glycogen by consuming a diet high in carbohydrates. Meeting energy needs can help improve performance during the sport, as well as improve overall strength and endurance.

There are different kinds of carbohydrates—simple or refined, and unrefined. A typical American consumes about 50% of their carbohydrates as simple sugars, which are added to foods as opposed to sugars that come naturally in fruits and vegetables.

These simple sugars come in large amounts in sodas and fast food. Over the course of a year, the average American consumes 54 gallons of soft drinks, which contain the highest amount of added sugars. Even though carbohydrates are necessary for humans to function, they are not all equally healthful. When machinery has been used to remove bits of high fiber, the carbohydrates are refined. These are the carbohydrates found in white bread and fast food.

Malnutrition

Malnutrition refers to insufficient, excessive, or imbalanced consumption of nutrients. In developed countries, the diseases of malnutrition are most often associated with nutritional imbalances or excessive consumption. Although there are more people in the world who are malnourished due to excessive consumption, according to the United Nations World Health Organization, the real challenge in developing nations today, more than starvation, is combating insufficient nutrition — the lack of nutrients necessary for the growth and maintenance of vital functions.

Mental Agility

Research indicates that improving the awareness of nutritious meal choices and establishing long-term habits of healthy eating have a positive effect on cognitive and spatial memory capacity, potentially increasing a student's potential to process and retain academic information. Some organizations have begun working with teachers, policymakers, and managed foodservice contractors to mandate improved nutritional content and increased nutritional resources in school cafeterias from primary to university level institutions. Health and nutrition have been proven to have close links with overall educational success. Currently, less than 10% of American college students report that they eat the recommended five servings of fruit and vegetables daily. Better nutrition has been shown to have an impact on both cognitive and spatial memory performance; a study showed those with higher blood sugar levels performed better on certain memory tests. In another study, those who consumed yogurt performed better on thinking tasks when compared to those who consumed caffeine free diet soda or confections. Nutritional deficiencies have been shown to have a negative effect on learning behaviour in mice as far back as 1951.

"Better learning performance is associated with diet induced effects on learning and memory ability". The "nutrition-learning nexus" demonstrates the correlation between diet and learning and has application in a higher education setting.

"We find that better nourished children perform significantly better in school, partly because they enter school earlier and thus have more time to learn but mostly because of greater learning productivity per year of schooling."

91% of college students feel that they are in good health while only 7% eat their recommended daily allowance of fruits and vegetables.

Nutritional education is an effective and workable model in a higher education setting. More "engaged" learning models that encompass nutrition is an idea that is picking up steam at all levels of the learning cycle.

There is limited research available that directly links a student's Grade Point Average (G.P.A.) to their overall nutritional health. Additional substantive data is needed to prove that overall intellectual health is closely linked to a person's diet, rather than just another correlation fallacy.

Mental Disorders

Nutritional supplement treatment may be appropriate for major depression, bipolar disorder, schizophrenia, and obsessive compulsive disorder, the four most common mental disorders in developed countries. Supplements that have been studied most for mood elevation and stabilization include eicosapentaenoic acid and docosahexaenoic acid (each of which are an omega-3 fatty acid contained in fish oil, but not in flaxseed oil), vitamin B_{12}, folic acid, and inositol.

Cancer

Cancer is now common in developing countries. According to a study by the International Agency for Research on Cancer, "In the developing world, cancers of the liver, stomach and esophagus were more common, often linked to consumption of carcinogenic preserved foods, such as smoked or salted food, and parasitic infections that attack organs." Lung cancer rates are rising rapidly in poorer nations because of increased use of tobacco. Developed countries "tended to have cancers linked to affluence or a 'Western lifestyle' — cancers of the colon, rectum, breast and prostate — that can be caused by obesity, lack of exercise, diet and age."

Metabolic Syndrome

Several lines of evidence indicate lifestyle-induced hyperinsulinemia and reduced insulin function (i.e. insulin resistance) as a decisive factor in many disease states. For example, hyperinsulinemia and insulin resistance are strongly linked to chronic inflammation, which in turn is strongly linked to a variety of adverse developments such as arterial microinjuries and clot formation (i.e. heart disease) and exaggerated cell division (i.e. cancer). Hyperinsulinemia and insulin resistance (the so-called metabolic syndrome) are characterized by a combination of abdominal obesity, elevated blood sugar, elevated blood pressure, elevated blood triglycerides, and reduced HDL cholesterol. The negative impact of hyperinsulinemia on prostaglandin PGE1/PGE2 balance may be significant.

The state of obesity clearly contributes to insulin resistance, which in turn can cause type 2 diabetes. Virtually all obese and most type 2 diabetic individuals have marked insulin resistance. Although the association between overweight and insulin resistance is clear, the exact (likely multifarious) causes of insulin resistance remain less clear. Importantly, it has been demonstrated that appropriate exercise, more regular food intake and reducing glycemic load all can reverse insulin resistance in overweight individuals (and thereby lower blood sugar levels in those who have type 2 diabetes).

Obesity can unfavourably alter hormonal and metabolic status via resistance to the hormone leptin, and a vicious cycle may occur in which insulin/leptin resistance and obesity aggravate one another.

The vicious cycle is putatively fuelled by continuously high insulin/leptin stimulation and fat storage, as a result of high intake of strongly insulin/leptin stimulating foods and energy. Both insulin and leptin normally function as satiety signals to the hypothalamus in the brain; however, insulin/leptin resistance may reduce this signal and therefore allow continued overfeeding despite large body fat stores. In addition, reduced leptin signalling to the brain may reduce leptin's normal effect to maintain an appropriately high metabolic rate.

There is a debate about how and to what extent different dietary factors— such as intake of processed carbohydrates, total protein, fat, and carbohydrate intake, intake of saturated and trans fatty acids, and low intake of vitamins/minerals—contribute to the development of insulin and leptin resistance. In any case, analogous to the way modern man-made pollution may potentially overwhelm the environment's ability to maintain homeostasis, the recent explosive introduction of high glycemic index and processed foods into the human diet may potentially overwhelm the body's ability to maintain homeostasis and health (as evidenced by the metabolic syndrome epidemic).

Hyponatremia

Excess water intake, without replenishment of sodium and potassium salts, leads to hyponatremia, which can further lead to water intoxication at more dangerous levels. A well-publicized case occurred in 2007, when Jennifer Strange died while participating in a water-drinking contest. More usually, the condition occurs in long-distance endurance events (such as marathon or triathlon competition and training) and causes gradual mental dulling, headache, drowsiness, weakness, and confusion; extreme cases may result in coma, convulsions, and death. The primary damage comes from swelling of the brain, caused by increased osmosis as blood salinity decreases.

Antinutrient

Antinutrients are natural or synthetic compounds that interfere with the absorption of nutrients. Nutrition studies focus on antinutrients commonly found in food sources and beverages.

Processed Foods

Since the Industrial Revolution some two hundred years ago, the food processing industry has invented many technologies that both help keep foods fresh longer and alter the fresh state of food as they appear in nature. Cooling is the primary technology used to maintain freshness, whereas many more technologies have been invented to allow foods to last longer without becoming spoiled. These latter technologies include pasteurisation, autoclavation, drying, salting, and separation of various components, all of which appear to alter the original nutritional contents of food. Pasteurisation and autoclavation (heating techniques) have no doubt improved the safety of many common foods, preventing epidemics of bacterial infection. But some of the (new) food processing technologies undoubtedly have downfalls as well.

Modern separation techniques such as milling, centrifugation, and pressing have enabled concentration of particular components of food, yielding flour, oils, juices and so on, and even separate fatty acids, amino acids, vitamins, and minerals. Inevitably, such large scale concentration changes the nutritional content of food, saving certain nutrients while removing others. Heating techniques may also reduce food's content of many heat-labile nutrients such as certain vitamins and phytochemicals, and possibly other yet to be discovered substances.

Because of reduced nutritional value, processed foods are often 'enriched' or 'fortified' with some of the most critical nutrients (usually certain vitamins) that were lost during processing. Nonetheless, processed foods tend to have an inferior nutritional profile compared to whole, fresh foods, regarding content of both sugar and high GI starches, potassium/sodium, vitamins, fiber, and of intact, unoxidized (essential) fatty acids. In addition, processed foods often contain potentially harmful substances such as oxidized fats and trans fatty acids.

A dramatic example of the effect of food processing on a population's health is the history of epidemics of beri-beri in people subsisting on polished rice. Removing the outer layer of rice by polishing it removes with it the essential vitamin thiamine, causing beri-beri. Another example is the development of scurvy among infants in the late 19th century in the United States. It turned out that the vast majority of sufferers were being fed milk that had been heat-treated (as suggested by Pasteur) to control bacterial disease. Pasteurisation was effective against bacteria, but it destroyed the vitamin C.

As mentioned, lifestyle-and obesity-related diseases are becoming increasingly prevalent all around the world. There is little doubt that the increasingly widespread application of some modern food processing technologies has contributed to this development.

The food processing industry is a major part of modern economy, and as such it is influential in political decisions (e.g. nutritional recommendations, agricultural subsidising). In any known profit-driven economy, health considerations are hardly a priority; effective production of cheap foods with a long shelf-life is more the trend. In general, whole, fresh foods have a relatively short shelf-life and are less profitable to produce and sell than are more processed foods.

Thus, the consumer is left with the choice between more expensive, but nutritionally superior, whole, fresh foods, and cheap, usually nutritionally inferior, processed foods. Because processed foods are often cheaper, more convenient (in both purchasing, storage, and preparation), and more available, the consumption of nutritionally inferior foods has been increasing throughout the world along with many nutrition-related health complications.

History

Humans have evolved as omnivorous hunter-gatherers over the past 250,000 years. The diet of early modern humans varied significantly depending on location and climate. The diet in the tropics tended to be based more heavily on plant foods, while the diet at higher latitudes tended more towards animal products. Analysis of postcranial and cranial remains of humans and animals from the Neolithic, along with detailed bone modification studies have shown that cannibalism was also prevalent among prehistoric humans.

Agriculture developed about 10,000 years ago in multiple locations throughout the world, providing grains such as wheat, rice, potatoes, and maize, with staples such as bread, pasta, and tortillas. Farming also provided milk and dairy products, and sharply increased the availability of meats and the diversity of vegetables. The importance of food purity was recognized when bulk storage led to infestation and contamination risks. Cooking developed as an often ritualistic activity, due to efficiency and reliability concerns requiring adherence to strict recipes and procedures, and in response to demands for food purity and consistency.

From Antiquity to 1900

The first recorded nutritional experiment is found in the Bible's Book of Daniel. Daniel and his friends were captured by the king of Babylon during an invasion of Israel. Selected as court servants, they were to share in the king's fine foods and wine. But they objected, preferring vegetables (pulses) and water in accordance with their

Jewish dietary restrictions. The king's chief steward reluctantly agreed to a trial. Daniel and his friends received their diet for 10 days and were then compared to the king's men. Appearing healthier, they were allowed to continue with their diet.

Around 475 BC, Anaxagoras stated that food is absorbed by the human body and therefore contained "homeomerics" (generative components), suggesting the existence of nutrients. Around 400 BC, Hippocrates said, "Let food be your medicine and medicine be your food."

In the 16th century, scientist and artist Leonardo da Vinci compared metabolism to a burning candle. In 1747, Dr. James Lind, a physician in the British navy, performed the first scientific nutrition experiment, discovering that lime juice saved sailors who had been at sea for years from scurvy, a deadly and painful bleeding disorder. The discovery was ignored for forty years, after which British sailors became known as "limeys." The essential vitamin C within lime juice would not be identified by scientists until the 1930s.

Around 1770, Antoine Lavoisier, the "Father of Nutrition and Chemistry" discovered the details of metabolism, demonstrating that the oxidation of food is the source of body heat. In 1790, George Fordyce recognized calcium as necessary for fowl survival. In the early 19th century, the elements carbon, nitrogen, hydrogen and oxygen were recognized as the primary components of food, and methods to measure their proportions were developed.

In 1816, François Magendie discovered that dogs fed only carbohydrates and fat lost their body protein and died in a few weeks, but dogs also fed protein survived, identifying protein as an essential dietary component. In 1840, Justus Liebig discovered the chemical makeup of carbohydrates (sugars), fats (fatty acids) and proteins (amino acids.) In the 1860s, Claude Bernard discovered that body fat can be synthesized from carbohydrate and protein, showing that the energy in blood glucose can be stored as fat or as glycogen.

In the early 1880s, Kanehiro Takaki observed that Japanese sailors (whose diets consisted almost entirely of white rice) developed beriberi (or endemic neuritis, a disease causing heart problems and paralysis), but British sailors and Japanese naval officers did not. Adding various types of vegetables and meats to the diets of Japanese sailors prevented the disease. In 1896, Eugen Baumann observed iodine in thyroid glands.

In 1897, Christiaan Eijkman worked with natives of Java, who also suffered from beriberi. Eijkman observed that chickens fed the native diet of white rice developed the symptoms of beriberi, but remained healthy when fed unprocessed brown rice with the outer bran intact. Eijkman cured the natives by feeding them brown rice, discovering that food can cure disease. Over two decades later, nutritionists learned that the outer rice bran contains vitamin B1, also known as thiamine.

From 1900 to the Present

In the early 20th century, Carl Von Voit and Max Rubner independently measured caloric energy expenditure in different species of animals, applying principles of physics in nutrition. In 1906, Wilcock and Hopkins showed that the amino acid tryptophan was necessary for the survival of rats. He fed them a special mixture of food containing all the nutrients he believed were essential for survival, but the rats died. A second group of rats were fed an amount of milk containing vitamins. Gowland Hopkins recognized "accessory food factors" other than calories, protein and minerals, as organic materials essential to health, but which the body cannot synthesize. In 1907, Stephen M. Babcock and Edwin B. Hart conducted the single-grain experiment, which took nearly four years to complete.

In 1912, Casimir Funk coined the term vitamin, a vital factor in the diet, from the words "vital" and "amine," because these unknown substances preventing scurvy, beriberi, and pellagra, were thought then to be derived from ammonia. The vitamins were studied in the first half of the 20th century.

In 1913, Elmer McCollum discovered the first vitamins, fat soluble vitamin A, and water soluble vitamin B (in 1915; now known to be a complex of several water-soluble vitamins) and named vitamin C as the then-unknown substance preventing scurvy.

Lafayette Mendel and Thomas Osborne also performed pioneering work on vitamins A and B. In 1919, Sir Edward Mellanby incorrectly identified rickets as a vitamin A deficiency because he could cure it in dogs with cod liver oil. In 1922, Elmer McCollum destroyed the vitamin A in cod liver oil, but found that it still cured rickets. Also in 1922, H.M. Evans and L.S. Bishop discover vitamin E as essential for rat pregnancy, originally calling it "food factor X" until 1925.

In 1925, Hart discovered that trace amounts of copper are necessary for iron absorption. In 1927, Adolf Otto Reinhold Windaus synthesized vitamin D, for which he won the Nobel Prize in Chemistry in 1928. In 1928, Albert Szent-Györgyi isolated ascorbic acid, and in 1932 proved that it is vitamin C by preventing scurvy. In 1935 he synthesized it, and in 1937, he won a Nobel Prize for his efforts. Szent-Györgyi concurrently elucidated much of the citric acid cycle.

In the 1930s, William Cumming Rose identified essential amino acids, necessary protein components which the body cannot synthesize. In 1935, Underwood and Marston independently discovered the necessity of cobalt. In 1936, Eugene Floyd Dubois showed that work and school performance are related to caloric intake. In 1938, Erhard Fernholz discovered the chemical structure of vitamin E. It was synthesised by Paul Karrer.

In 1940, rationing in the United Kingdom during and after World War II took place according to nutritional principles drawn up by Elsie Widdowson and others. In 1941, the first Recommended Dietary Allowances (RDAs) were established by the National Research Council.

In 1992, The U.S. Department of Agriculture introduced the Food Guide Pyramid. In 2002, a Natural Justice study showed a relation between nutrition and violent behaviour. In 2005, a study found that obesity may be caused by adenovirus in addition to bad nutrition.

Plant Nutrition

Plant nutrition is the study of the chemical elements that are necessary for plant growth. There are several principles that apply to plant nutrition. Some elements are directly involved in plant metabolism. However, this principle does not account for the so-called beneficial elements, whose presence, while not required, has clear positive effects on plant growth. A nutrient that is able to limit plant growth according to Liebig's law of the minimum, is considered an essential plant nutrient if the plant cannot complete its full life cycle without it. There are 17 essential plant nutrients.

Macronutrients:

- N = Nitrogen
- P = Phosphorus
- K = Potassium
- Ca = Calcium
- Mg = Magnesium
- S = Sulfur
- Si = Silicon.

Micronutrients (trace levels) include:

- Cl = Chlorine
- Fe = Iron
- B = Boron
- Mn = Manganese
- Na = Sodium
- Zn = Zinc
- Cu = Copper
- Ni= Nickel
- Mo = Molybdenum.

Macronutrients

Calcium

Calcium regulates transport of other nutrients into the plant and is also involved in the activation of certain plant enzymes. Calcium deficiency results in stunting.

Nitrogen

Nitrogen is an essential component of all proteins. Nitrogen deficiency most often results in stunted growth.

Phosphorus

Phosphorus is important in plant bioenergetics. As a component of ATP, phosphorus is needed for the conversion of light energy to chemical energy (ATP) during photosynthesis. Phosphorus can also be used to modify the activity of various enzymes by phosphorylation, and can be used for cell signalling. Since ATP can be used for the biosynthesis of many plant biomolecules, phosphorus is important for plant growth and flower/seed formation.

Potassium

Potassium regulates the opening and closing of the stoma by a potassium ion pump. Since stomata are important in water regulation, potassium reduces water loss from the leaves and increases drought tolerance. Potassium deficiency may cause necrosis or interveinal chlorosis.

Silicon

Silicon is deposited in cell walls and contributes to its mechanical properties including rigidity and elasticity

Micronutrients

Boron

Boron is important in sugar transport, cell division, and synthesizing certain enzymes. Boron deficiency causes necrosis in young leaves and stunting.

Copper

Copper is important for photosynthesis. Symptoms for copper deficiency include chlorosis. Involved in many enzyme processes. Necessary for proper photosythesis. Involved in the manufacture of lignin (cell walls). Involved in grain production.

Chlorine

Chlorine is necessary for osmosis and ionic balance; it also plays a role in photosynthesis.

Iron

Iron is necessary for photosynthesis and is present as an enzyme cofactor in plants. Iron deficiency can result in interveinal chlorosis and necrosis.

Manganese

Manganese is necessary for building the chloroplasts. Manganese deficiency may result in coloration abnormalities, such as discolored spots on the foliage.

Molybdenum

Molybdenum is a cofactor to enzymes important in building amino acids.

Nickel

In higher plants, Nickel is essential for activation of urease, an enzyme involved with nitrogen metabolism that is required to process urea. Without Nickel, toxic levels of urea accumulate, leading to the formation of necrotic lesions. In lower plants, Nickel activates several enzymes involved in a variety of processes, and can substitute for Zinc and Iron as a cofactor in some enzymes.

Sodium

Sodium is involved in the regeneration of phosphoenolpyruvate in CAM and C4 plants. It can also substitute for potassium in some circumstances.

Zinc

Zinc is required in a large number of enzymes and plays an essential role in DNA transcription. A typical symptom of zinc deficiency is the stunted growth of leaves, commonly known as "little leaf" and is caused by the oxidative degradation of the growth hormone auxin.

Processes

Plants uptake essential elements from the soil through their roots and from the air (mainly consisting of nitrogen and oxygen) through their leaves. Nutrient uptake in the soil is achieved by cation exchange, wherein root hairs pump hydrogen ions (H^+) into the soil through proton pumps. These hydrogen ions displace cations attached to negatively charged soil particles so that the cations are available for uptake by the root. In the leaves, stomata open to take in carbon dioxide and expel oxygen.

The carbon dioxide molecules are used as the carbon source in photosynthesis. Although nitrogen is plentiful in the Earth's atmosphere, relatively few plants engage in nitrogen fixation (conversion of atmospheric nitrogen to a biologically useful form). Most plants therefore require nitrogen compounds to be present in the soil in which they grow.

Plant nutrition is a difficult subject to understand completely, partially because of the variation between different plants and even between different species or individuals of a given clone.

Elements present at low levels may cause deficiency symptoms, and toxicity is possible at levels that are too high. Furthermore, deficiency of one element may present as symptoms of toxicity from another element, and vice-versa.

Carbon and oxygen are absorbed from the air, while other nutrients are absorbed from the soil. Green plants obtain their carbohydrate supply from the carbon dioxide in the air by the process of photosynthesis.

Malnutrition

Malnutrition refers to insufficient, excessive, or imbalanced consumption of nutrients by an organism. In developed countries, the diseases of malnutrition are most often associated with nutritional imbalances or excessive consumption.

Although there are more organisms in the world who are malnourished due to insufficient consumption, increasingly more organisms suffer from excessive over-nutrition; a problem caused by an over abundance of sustenance coupled with the instinctual desire (by animals in particular) to consume all that it can.

Insufficient

Under consumption generally refers to the long-term consumption of insufficient sustenance in relation to the energy that an organism expends or expels, leading to poor health.

Excessive

Over consumption generally refers to the long-term consumption of excess sustenance in relation to the energy that an organism expends or expels, leading to poor health and, in animals, obesity. It can cause excessive hair loss, brittle nails, and irregular premenstrual cycles for females

Unbalanced

When too much of one or more nutrients is present in the diet to the exclusion of the proper amount of other nutrients, the diet is said to be unbalanced.

Nutritional biochemistry is the study of nutrition as a science. While nutritional science is composed of various studies in food components, nutrients, and their function regarding humans and other mammals, nutritional biochemistry specifically focuses on nutrient chemical components, and how they function metabolically, physiologically, and biochemically.

Biochemistry research in this field is primarily centered upon defining dietary regulations for the general public.

An integrative form of science, nutritional biochemistry uses other sciences, such as chemistry, biology, and physics. It applies these sciences specifically to the study of health, diet, disease, and the connections that exist among them. To do this, biochemists might ascertain the structures of different nutrients, or use a biochemistry lab to discover how specific nutrients might affect the metabolism. Such research can lead to the development of public health policies and interventions.

Malnutrition is typically a serious concern for a nutritional biochemist. It is often the cause of disabilities, disease, and other problems due to nutritional deficiencies or overnutrition.

These conditions can include diabetes, cancer, birth defects, obesity, and cardiovascular disease. Through the study of nutritional biochemistry, scientists can develop plans to counteract these afflictions, identifying both nutritional causes as well as preventative measures.

Genetics are also commonly studied by nutritional biochemists. Many diseases and dietary factors are affected by genetics; biochemists can use genetic studies to pinpoint these different effects, and improve public health overall.

The fortification of many different food products, promotion of low-cholesterol diets, and iodinization of table salt were all introduced into society through recommendations made through nutritional biochemistry findings.

Probiotic

Probiotics are live microorganisms thought to be healthy for the host organism. According to the currently adopted definition by FAO/ WHO, probiotics are: "Live microorganisms which when administered in adequate amounts confer a health benefit on the host". Lactic acid bacteria (LAB) and bifidobacteria are the most common types of microbes used as probiotics; but certain yeasts and bacilli may also be helpful. Probiotics are commonly consumed as part of fermented foods with specially added active live cultures; such as in yogurt, soy yogurt, or as dietary supplements.

At the start of the 20th century, probiotics were thought to beneficially affect the host by improving its intestinal microbial balance, thus inhibiting pathogens and toxin producing bacteria. Today, specific health effects are being investigated and documented including alleviation of chronic intestinal inflammatory diseases, prevention and treatment of pathogen-induced diarrhea, urogenital infections, and atopic diseases.

History of Probiotics

The original observation of the positive role played by certain bacteria was first introduced by Russian scientist and Nobel laureate Eli Metchnikoff, who in the beginning of the 20th century suggested that it would be possible to modify the gut flora and to replace harmful microbes with useful microbes. Metchnikoff, at that time a professor at the Pasteur Institute in Paris, produced the notion that the aging process results from the activity of putrefactive (proteolytic) microbes producing toxic substances in the large bowel. Proteolytic bacteria such as clostridia, which are part of the normal gut flora, produce toxic substances including phenols, indols and ammonia from the digestion of proteins. According to Metchnikoff these compounds were responsible for what he called "intestinal auto-intoxication", which caused the physical changes associated with old age.

It was at that time known that milk fermented with lactic-acid bacteria inhibits the growth of proteolytic bacteria because of the low pH produced by the fermentation of lactose. Metchnikoff had also observed that certain rural populations in Europe, for example in Bulgaria and the Russian steppes who lived largely on milk fermented by lactic-acid bacteria were exceptionally long lived. Based on these facts, Metchnikoff proposed that consumption of fermented milk would "seed" the intestine with harmless lactic-acid bacteria and decrease the intestinal pH and that this would suppress the growth of proteolytic bacteria. Metchnikoff himself introduced in his diet sour milk fermented with the bacteria he called "Bulgarian Bacillus" and found his health benefited. Friends in Paris soon followed his example and physicians began prescribing the sour milk diet for their patients.

Bifidobacteria were first isolated from a breast-fed infant by Henry Tissier who also worked at the Pasteur Institute. The isolated bacterium named Bacillus bifidus communis was later renamed to the genus *Bifidobacterium*. Tissier found that bifidobacteria are dominant in the gut flora of breast-fed babies and he observed clinical benefits from treating diarrhea in infants with bifidobacteria. The claimed effect was bifidobacterial displacement of proteolytic bacteria causing the disease. During an outbreak of shigellosis in 1917, German professor Alfred Nissle isolated a strain of *Escherichia coli* from the feces of a soldier who was not affected by the disease. Methods of treating infectious diseases were needed at that time when antibiotics were not yet available, and Nissle used the *Escherichia coli* Nissle 1917 strain in acute gastrointestinal infectious salmonellosis and shigellosis.

In 1920, Rettger demonstrated that Metchnikoff's "Bulgarian Bacillus", later called *Lactobacillus delbrueckii subsp. bulgaricus*, could not live in the human intestine, and the fermented food phenomena petered out. Metchnikoff's theory was disputable (at this stage), and people doubted his theory of longevity.

After Metchnikoff's death in 1916, the centre of activity moved to the United States. It was reasoned that bacteria originating from the gut were more likely to produce the desired effect in the gut, and in 1935 certain strains of *Lactobacillus acidophilus* were found to be very active when implanted in the human digestive tract. Trials were carried out using this organism, and encouraging results were obtained especially in the relief of chronic constipation.

The term "probiotics" was first introduced in 1953 by Kollath. Contrasting antibiotics, probiotics were defined as microbially derived factors that stimulate the growth of other microorganisms. In 1989 Roy Fuller suggested a definition of probiotics which has been widely used: "*A live microbial feed supplement which beneficially affects the host animal by improving its intestinal microbial balance*". Fuller's definition emphasizes the requirement of viability for probiotics and introduces the aspect of a beneficial effect on the host.

In the following decades intestinal lactic acid bacterial species with alleged health beneficial properties have been introduced as probiotics, including *Lactobacillus rhamnosus*, *Lactobacillus casei*, and *Lactobacillus johnsonii*.

Potential Benefits

Experiments into the benefits of probiotic therapies suggest a range of potentially beneficial medicinal uses for probiotics. For many of the potential benefits, research is limited and only preliminary results are available. It should be noted that the effects described are *not* general effects of probiotics. Recent research on the molecular biology and genomics of *Lactobacillus* has focused on the interaction with the immune system, anti-cancer potential, and potential as a biotherapeutic agent in cases of antibiotic-associated diarrhoea, travellers' diarrhoea, pediatric diarrhoea, inflammatory bowel disease and irritable bowel syndrome.

All effects can only be attributed to the individual strain(s) tested. Testing of a supplement does not indicate benefit from any other strain of the same species, and testing does not indicate benefit from the whole group of LAB (or other probiotics).

Managing Lactose Intolerance

As lactic acid bacteria actively convert lactose into lactic acid, ingestion of certain active strains may help lactose intolerant individuals tolerate more lactose than what they would have otherwise.

Prevention of Colon Cancer

In laboratory investigations, some strains of LAB (*Lactobacillus bulgaricus*) have demonstrated anti-mutagenic effects thought to be due to their ability to bind with heterocyclic amines, which are carcinogenic substances formed in cooked meat. Animal studies have demonstrated that some LAB can protect against colon cancer in rodents, though human data is limited and conflicting. Most human trials have found that the strains tested may exert anti-carcinogenic effects by decreasing the activity of an enzyme called b-glucuronidase (which can generate carcinogens in the digestive system). Lower rates of colon cancer among higher consumers of fermented dairy products have been observed in one population study.

Lowering Cholesterol

Animal studies have demonstrated the efficacy of a range of LAB to be able to lower serum cholesterol levels, presumably by breaking down bile in the gut, thus inhibiting its reabsorption (which enters the blood as cholesterol). Some, but not all human trials have shown that dairy foods fermented with specific LAB can produce modest reductions in total and LDL cholesterol levels in those with normal levels to begin with, however trials in hyperlipidemic subjects are needed.

Lowering Blood Pressure

Several small clinical trials have indicated that consumption of milk fermented with various strains of LAB may result in modest reductions in blood pressure. It is thought that this is due to the ACE inhibitor-like peptides produced during fermentation.

Improving Immune Function and Preventing Infections

LAB are thought to have several presumably beneficial effects on immune function. They may protect against pathogens by means of competitive inhibition (i.e., by competing for growth) and there is evidence to suggest that they may improve immune function by increasing the number of IgA-producing plasma cells, increasing or improving phagocytosis as well as increasing the proportion of T lymphocytes and Natural Killer cells. Clinical trials have demonstrated that probiotics may decrease the incidence of respiratory tract infections and dental caries in children. LAB foods and supplements have been shown to aid in the treatment and prevention of acute diarrhea, and in decreasing the severity and duration of rotavirus infections in children and travelers' diarrhea in adults.

A 2010 study suggested that the anecdotal benefits of probiotic therapies as beneficial for preventing secondary infections, a common complication of antibiotic therapy, may be because keeping the immune system primed by eating foods enhanced with "good" bacteria may help counteract the negative effects of sickness and antibiotics. It was thought that antibiotics may turn the immune system "off" while probiotics turns it back on "idle", and more able to quickly react to new infections.

Helicobacter Pylori

LAB are also thought to aid in the treatment of *Helicobacter pylori* infections (which cause peptic ulcers) in adults when used in combination with standard medical treatments. However more studies are required into this area.

Antibiotic-associated Diarrhea

Antibiotic-associated diarrhea (AAD) results from an imbalance in the colonic microbiota caused by antibiotic therapy. Microbiota alteration changes carbohydrate metabolism with decreased short-chain fatty acid absorption and an osmotic diarrhea as a result. Another consequence of antibiotic therapy leading to diarrhea is overgrowth of potentially pathogenic organisms such as *Clostridium difficile*.

Probiotic treatment can reduce the incidence and severity of AAD as indicated in several meta-analyses. However, further documentation of these findings through randomized, double blind, placebo-controlled trials are warranted.

Efficacy of probiotic AAD prevention is dependent on the probiotic strain(s) used and on the dosage. Up to a 50% reduction of AAD occurrence has been found. No side-effects have been reported in any of these studies. Caution should, however, be exercised when administering probiotic supplements to immunocompromised individuals or patients who have a compromised intestinal barrier.

Reducing Inflammation

LAB and supplements have been found to modulate inflammatory and hypersensitivity responses, an observation thought to be at least in part due to the regulation of cytokine function. Clinical studies suggest that they can prevent reoccurrences of inflammatory bowel disease in adults, as well as improve milk allergies. They are not effective for treating eczema, a persistent skin inflammation. How probiotics counteract immune system overactivity remains unclear, but a potential mechanism is desensitization of T lymphocytes, an important component of the immune system, towards pro-inflammatory stimuli.

Improving Mineral Absorption

It is hypothesized that probiotic lactobacilli may help correct malabsorption of trace minerals, found particularly in those with diets high in phytate content from whole grains, nuts, and legumes.

Irritable Bowel Syndrome and Colitis

B. infantis 35624, sold as Align, was found to improve some symptoms of irritable bowel syndrome in women in a recent study. Another probiotic bacterium, *Lactobacillus plantarum* 299v, was also found to be effective in reducing IBS symptoms.

Additionally, a probiotic formulation, VSL#3, was found to be safe in treating ulcerative colitis, though efficacy in the study was uncertain. *Bifidobacterium animalis* DN-173 010 may help. For maintenance of remission of ulcerative colitis, Mutaflor (*E.coli* Nissle 1917) there are 3 controlled, randomized, double blind clinical studies which have proven equivalence of Mutaflor and mesalazine (5-ASAs).

Managing Urogenital Health

Several in vitro studies have revealed probiotics' potential in relieving urinary tract infections and bacterial vaginosis. Results have been varied on these studies, and in vivo studies are still required in this area to determine efficacy.

Disadvantages of Probiotics

In some specific situations (such as critically ill patients) they could be harmful. In a therapeutic clinical trial conducted by the Dutch Pancreatitis Study Group, the consumption of a mixture of six probiotic bacteria, increased the death rate of patients with predicted severe acute pancreatitis.

In a clinical trial conducted at the University of Western Australia, aimed at showing the effectiveness of probiotics in reducing childhood allergies, Dr Susan Prescott and her colleagues gave 178 children either a probiotic or a placebo for the first six months of their life. Those given the good bacteria were more likely to develop a sensitivity to allergens. Some hospitals have reported treating lactobacillus septicaemia, which is a potentially fatal disease caused by the consumption of probiotics by people with lowered immune systems or who are already very ill.

There is no published evidence that probiotic supplements are able to replace the body's natural flora when these have been killed off; indeed bacterial levels in feces disappear within days when supplementation ceases.

Recent studies indicate that probiotic products such as yogurts could be a cause for obesity trends. However, this is contested.

Strains

Live probiotic cultures are available in fermented dairy products and probiotic fortified foods.

However, tablets, capsules, powders and sachets containing the bacteria in freeze dried form are also available.

Some other bacteria mentioned in probiotic products: (yoghurt bacteria):

- Lactobacillus bulgaricus
- Streptococcus thermophilus
- Lactobacillus bifidus-became new genus Bifidobacterium.

Some fermented products containing similar lactic acid bacteria include:

- Pickled vegetables
- Fermented bean paste such as tempeh, miso and doenjang
- Kefir
- Buttermilk or Karnemelk
- Kimchi
- Pao cai
- Sauerkraut
- Soy sauce
- Zha cai.

Multi-probiotic

Research is emerging on the potential health benefits of multiple probiotic strains as a health supplement as opposed to a single strain. The human gut is home to some 400-500 types of microbes.

It is thought that this diverse environment may benefit from multiple probiotic strains; different strains populate different areas of the digestive tract, and studies are beginning to link different probiotic strains to specific health benefits.

Synbiotics and Prebiotics

While related, prebiotics and synbiotics are not the same as probiotics.

Other Uses

Probiotics are used in industry to improve yields of pork and chicken production.

Research

Research about probiotics shows benefits.

A study in 2004 testing the immune system of students given either milk or Actimel over a 6 week exam period (3 weeks of studying, 3 weeks of exams) tested 19 different biomarkers. Of these 19 biomarkers only 2 were shown to be different between the two groups, increased production of lymphocytes and increased production of CD56 cells. The tests were not blind and show that certain probiotic strains may have no overall effect on the immune system or on its ability.

A 2007 study at University College Cork in Ireland showed that a diet including milk fermented with *Lactobacillus* bacteria prevented *Salmonella* infection in pigs.

A 2007 clinical study at Imperial College London showed that preventive consumption of a commercially available probiotic drink containing *L casei DN-114001*, *L bulgaricus*, and *S thermophilus* can reduce the incidence of antibiotic-associated diarrhea and *C difficile*-associated diarrhea.

The efficacy and safety of a daily dose of *Lactobacillus acidophilus* CL1285 in the prevention of AAD was demonstrated by Montreal's Maisonneuve-Rosemont Hospital, in a clinical study of hospitalized patients.

Current research is focusing on the molecular biology and genomics of *Lactobacillus* and bifidobacteria. The application of modern whole genome approaches is providing insights into bifidobacterial evolution, while also revealing genetic functions that explain their presence in the particular ecological environment of the gastrointestinal tract.

Prebiotic Nutrition

Prebiotics are non-digestible food ingredients that stimulate the growth and/or activity of bacteria in the digestive system which are beneficial to the health of the body. They were first identified and named by Marcel Roberfroid in 1995. They are considered a functional food.

Typically, prebiotics are carbohydrates (such as oligosaccharides), but the definition may include non-carbohydrates. The most prevalent forms of prebiotics are nutritionally classed as soluble fiber. To some extent, many forms of dietary fiber exhibit some level of prebiotic effect. Roberfroid offered a refined definition in the 2007 Journal of Nutrition stating:

"A prebiotic is a selectively fermented ingredient that allows specific changes, both in the composition and/or activity in the gastrointestinal microflora that confers benefits upon host well-being and health." Additionally, in his 2007 revisit of Prebiotics, Roberfroid clarified that only two particular fructooligosaccharides fully meet this definition: oligofructose and inulin. Caution may be in order when referring to other substances as "prebiotic," and it is increasingly common to distinguish other similar substances by using terms such as "possible" or "likely" prebiotics. Many authorities also classify galactooligosaccharides (GOS) as "true" prebiotics. GOS is an animal-derived substance manufactured by extracting milk sugars from dairy products and using enzymes to break them down.

Researchers now also focus on the distinction between short-chain, long-chain, and full-spectrum prebiotics. "short-chain" prebiotics, e.g. oligofructose, contain 2-8 links per saccharide molecule, are typically fermented more quickly in the right-side of the colon providing nourishment to the bacteria in that area. Longer-chain prebiotics, e.g. Inulin, contain 9-64 links per saccharide molecule, and tend to be fermented more slowly, nourishing bacteria predominantly in the left-side colon. Full-spectrum prebiotics provide the full range of molecular link-lengths from 2-64 links per molecule, and nourish bacteria throughout the colon, e.g. Oligofructose-Enriched Inulin (OEI). The majority of research done on prebiotics is based on full-spectrum prebiotics, typically using OEI as the research substance.

Function

The prebiotic definition does not emphasize a specific bacterial group. Generally, however, it is assumed that a prebiotic should increase the number and/or activity of bifidobacteria and lactic acid bacteria. The importance of the bifidobacteria and the lactic acid bacteria (AKA lactobacillus or LABs) is that these groups of bacteria have several beneficial effects on the host, especially in terms of improving digestion (including enhancing mineral absorption) and the effectiveness and intrinsic strength of the immune system. A product that stimulates bifidobacteria is considered a bifidogenic factor. Some prebiotics may thus also act as a bifidogenic factor and vice versa, but the two concepts are not identical.

Sources

Traditional dietary sources of prebiotics include soybeans, inulin sources (such as Jerusalem artichoke, jicama, and chicory root), raw oats, unrefined wheat, unrefined barley and yacon. Some of the oligosaccharides that naturally occur in breast milk are believed to play an important role in the development of a healthy immune system in infants. It is becoming more common to properly distinguish between prebiotic substances and the food that contains them.

References to almonds, honey and other foods (most commonly in promotional materials from growers of those foods) as "a prebiotic" are not accurate. No plant or food *is* a prebiotic: Wheat, honey and many other foods *contain* prebiotics to a greater or lesser extent, ranging from fairly large portions (chicory root, Jerusalem artichoke) to only trace quantities (thousands of other plant-based foods). Referring to a food as "a prebiotic" is no more accurate than calling a food "a vitamin."

Table: *Top 10 Foods Containing Prebiotics*

Food	***Prebiotic Fiber Content by Weight***
Raw Chicory Root	64.6%
Raw Jerusalem Artichoke	31.5%
Raw Dandelion Greens	24.3%
Raw Garlic	17.5%
Raw Leek	11.7%
Raw Onion	8.6%
Cooked Onion	5%
Raw Asparagus	5%
Raw Wheat bran	5%
Whole Wheat flour, Cooked	4.8%
Raw Banana	1%

While there is no broad consensus on an ideal daily serving of prebiotics, recommendations typically range from 4-8g for general digestive health support, to 15g or more for those with active digestive disorders. Given an average 6g serving, below are the amounts of prebiotic foods required to achieve a daily serving of prebiotic fiber:

Food	***Amount of food to achieve 6g serving of prebiotics***
Raw Chicory Root	9.3g (about 1/3 oz)
Raw Jerusalem Artichoke	19g (about 3/4 oz)
Raw Dandelion Greens	24.7g (just under 1 oz)
Raw Garlic	34.3 g (about 1.2 oz)
Raw Leek	51.3g (about 1.8 oz)

Contd...

Food	***Amount of food to achieve 6g serving of prebiotics***
Raw Onion	69.8g (about 2.5 oz)
Cooked Onion	120g (about 1/4 lb)
Raw Asparagus	120g (about 1/4 lb)
Raw Wheat Bran	120g (about 1/4 lb)
Whole Wheat Flour, Cooked	125g (about 1/4 lb)
Raw Banana	600g (about 1.3 lb)

Those wishing to ensure sufficient prebiotic intake should carefully consider the prebiotic content of their diet, as well as what caloric/nutritive load comes along with it: e.g. eating 1.3 pounds of bananas daily is likely to provide an excess of calories & sugars/carbohydrates to the diet. Prebiotic fiber supplements with minimal caloric/fat/sugar load are also available.

Prebiotic oligosaccharides are increasingly added to foods for their health benefits. Some oligosaccharides that are used in this manner are fructooligosaccharides (FOS), xylooligosaccharides (XOS), polydextrose and galactooligosaccharides (GOS). Some monosaccharides such as tagatose are also used sometimes as prebiotics.

In petfood also mannooligosaccharides are being used for prebiotic purposes.

Genetically engineering plants for the production of inulins has also become more prevalent., despite the still limited insight into the immunological mechanisms activated by such food supplementation.

Effects

Studies have demonstrated positive effects on calcium and other mineral absorption, immune system effectiveness, bowel pH, reduction of colorectal cancer risk, inflammatory bowel disorders (Crohn's Disease and Ulcerative Colitis) Hypertension (high blood pressure) and intestinal regularity. Recent human trials have reinforced the role of Prebiotics in preventing and possibly stopping early stage colon cancer.

It has been argued that many of these health effects emanate from increased production of short-chain fatty acids (SCFA) by the stimulated beneficial bacteria. Thus food supplements specifically enhancing the growth of SCFA producing intestinal bacteria (such as clostridia and bacteroides species) are widely recognized to be beneficial.

While research does clearly demonstrate that prebiotics lead to increased production of these SCFA's, more research is required to establish a direct causal connection. It has been argued that prebiotics are beneficial to Crohn's Disease through production of SCFAs to nourish the colon walls, and beneficial to Ulcerative Colitis through reduction of Hydrogen Sulfide gas due to reduction of sulfate-reducing bacteria, which do not thrive in the slightly acidic environment SCFAs create.

The immediate addition of substantial quantities of prebiotics to the diet may result in a temporary increase in gas, bloating or bowel movement. It has been argued that chronically low consumption of prebiotic-containing foods in the typical Western diet may exaggerate this effect.

Human colonic bacteria substrates are relatively stable. Production of SCFA and fermentation quality are reduced during long-term diets of low fiber intake. Until bacterial flora are gradually established to habilitate or restore intestinal tone, nutrient absorption will be impaired and colonic transit time temporarily increased with an immediate addition of higher prebiotic intake.

Bibliography

Airey, D. and Johnson, S., : *The Profile of Tourism Studies Degree Courses in the UK: 1997/98,* London, 1998.

Ashraf, Syed Mahboob: *Handbook of Fruit and Vegetable Products*, Agrobios, Delhi, 2008.

Bagdi, Madan Lal: *Physiology, Biochemistry and Biotechnology,* Manglam Pub, Delhi, 2007.

Bainbridge, W.S.: *Societal Implications of Nanoscience and Nanotechnology*, Dordrecht, Kluwer, 2001.

Carroll, J.S.: *Social Science Research Methods for Assessing Societal Implications of Nanotechnology*, Kluwer, Dordrecht, 2001.

Chakraborty, Sudip : *Food Security and Child Labour : The Case of a Hazardous Occupation*, Deep and Deep, Delhi, 2011.

Daniel Ratner: *Nanotechnology: a Gentle Introduction to the Next Big Idea*, Prentice Hall, Upper Saddle River, 2003.

Daphne C. Elliott: *Biochemistry and Molecular Biology*, Oxford University Press, Delhi, 2005.

Eckel, Peter J.: *College & University Foodservice Management Standards*, Westport, AVI Pub. Company, 1985.

Elio, C.: *The Hospitality Law Desk Reference*, Miami, Southern Beverage Journal, 1994.

Featherly H. I.: *Taxonomic Terminology of the Higher Plants*, USA, Iowa State College Press, 1954.

Gautam Arvind : *A Concise Course on Hotel and Hospitality Management*, Axis Pub, Delhi, 2010.

Ghimire, Krishna: *The Native Tourist*: Mass Tourism within Developing Regions, London, Earthscan, 2001.

Hardy B.: *Biology and Agronomy of Forage Arachis*, International Centre for Tropical Agriculture, Cali, 1994.

Hays, Judi Radice: *Restaurant & Food Graphics*, Glen Cove, PBC International, 1994.

Ireland, Lewis: *Quality Management for Projects and Programs*, Upper Darby, PMI, 1991.

Jack Simons: *An Introduction to Theoretical Chemistry*, Cambridge Univ Press, Delhi, 1998.

Khurana A D : *Handbook of Catering, Hospitality and Tourism,* Mohit Pub, Delhi, 2008.

Kochhar B.K.: *Principles of Hospitality and Tourism Management*, Cyber Tech, Delhi, 2011.

Larkham, P J: *Building a New Heritage*: *Tourism, Culture & Identity in the New Europe*, London, Routledge,1994.

Laws, Eric: *Tourist Destination Management: Issues, Analysis & Policies*, London, Routledge, 1995.

Mayrose, V.B.: *Handbook of livestock Management Techniques*, Macmillan Publishing Co., New York, 1981.

Mowery, D. C. : *Technology and Wealth of Nations*, Stanford: Stanford University Press, 1992.

Nancy N.: *Choosing a Career in Hotels, Motels, and Resorts*, New York, Rosen Pub. Group, 1997.

Nobel, P. S.: *Physicochemical and Environmental Plant Physiology*, Academic Press, San Diego, 1999.

Ojha, Jai Shankar: *Aquaculture and Nutrition and Biochemistry,* Agrotech, Delhi, 2006.

Paul B. : *Food Biotechnology in Ethical Perspective*, Aspen, CO: Aspen Publishers, 1997.

Pearce, P.L., Moscardo, G.: *Tourism Community Relationships*, Oxford: Pergamon, 1996.

Ratti Manish*: Hospitality Management : Theories and Practices*, Rajat Pub, Delhi, 2007.

Robert C.: *Cases in Hospitality Marketing and Management*, New York, John Wiley, 1997.

Schummer, J.: *Interdisciplinary Issues in Nanoscale Research*, IOS Press, Amsterdam, 2004.

Thomas, F.: *Introduction to Management in the Hospitality Industry*, New York, Wiley, 1995.

Valli Manickam: *A Textbook of Analytical Chemistry*, PharmaMed Press, Delhi, 2006.

William H.: *The Chef's Guide to Practical Restaurant Cookery*, New York, Van Nostrand Reinhold, 1988.

Yadav, D.S.: *Advances in Vegetable Production*, Researchco Book Centre, Delhi, 2005.

Index

M

N

O

P

R

S

T

V

W

❑❑❑